T0205343

The Professional Development of Teachers: Practice and Theory

The Professional Development of Teachers: Practice and Theory

by

Philip Adey

King's College London,
United Kingdom

with

Gwen Hewitt, John Hewitt and Nicolette Landau

KLUWER ACADEMIC PUBLISHERS
DORDRECHT / BOSTON / LONDON

Library of Congress Cataloging-in-Publication Data

ISBN 1-4020-2005-8 (HB)
ISBN 1-4020-2006-6 (PB)

Published by Kluwer Academic Publishers,
P.O. Box 17, 3300 AA Dordrecht, The Netherlands.

Sold and distributed in North, Central and South America
by Kluwer Academic Publishers,
101 Philip Drive, Norwell, MA 02061, U.S.A.

In all other countries, sold and distributed
by Kluwer Academic Publishers,
P.O. Box 322, 3300 AH Dordrecht, The Netherlands.

Printed on acid-free paper

DEDICATION

for Jennifer Adey

… not just for her love and support for 40+ years, but for many valuable professional insights into the matter of this book from her experience as a headteacher, OfSTED inspector, and consultant.

CONTENTS

ABBREVIATIONS AND NOTES

CA   Cognitive Acceleration
CAME   Cognitive Acceleration through Mathematics Education
CASE   Cognitive Acceleration through Science Education
CATE   Cognitive Acceleration through Technology Education
CC   CASE Coordinator (Person in a school responsible for implementation of CASE)
DfES   Department for Education and Skills (the government ministry in England responsible for education)
GCSE   General Certificate of Secondary Education (national examination taken at end of Y11 in England and Wales)
HoD   Head of Department (in a school)
HoS   Head of Science (department in a school)
INSET   Inservice Education of Teachers
KS1 etc. Key Stage 1 etc. (see table below)
LEA   Local Education Authority (or 'Local Authority')
LoU   Level of Use (of an innovation)
NLS   National Literacy Strategy
NNS   National Numeracy Strategy
NQT   Newly Qualified Teacher (in their first year)
OfSTED   Office for Standards in Education (who inspect schools in England).
PD   Professional Development – in the context of this book, this refers generally to the continuing development of teachers after their initial training.
PKG   *Permatan Kerja Guru* – literally 'improving the work of teachers'.
Y1 etc   Year 1 etc. (see table below)
WISCIP   West Indian Science Curriculum Innovation Project
ZPD   Zone of Proximal Development

Ages, years, and grades in different systems

Age, years	5+	6+	7+	...	10+	11+	12+	13+	14+	15+	16+	17+
England	Y1	Y2	Y3	...	Y6	Y7	Y8	Y9	Y10	Y11	Y12	Y13
Key Stage	1		2			3			4		5	
School *	primary					secondary						
Scotland	P1	P2	P3	...	P6	P7	S1	S2	S3	S4	S5	
US grade	K	1	2	...	5	6	7	8	9	10	11	

* details vary widely across (and within) Local Authorities. For example, some have middle schools Y4 – Y7 or Y8.

PART 1: THE ISSUES AND SOME ATTEMPTED SOLUTIONS

1: INTRODUCTION

AN OUTLINE OF OUR AGENDA

We have subtitled this book 'Practice and Theory' because that is the order in which we plan to deal with the subject. We have been running and evaluating programmes for the professional development of teachers since 1970 and the first section of the book will describe some of that practice and the principles which have emerged, and then been re-cycled back into the practice. Key amongst those principles are:
- the necessarily long-term nature of inservice programmes which are to have a permanent effect on teaching practice;
- the central role of coaching work in schools; and
- the interaction between individual teacher factors and the department and school environment which encourages or discourages professional development.

Part 1 will describe some of the main professional development programmes for teachers with which we have been involved – in outline only for the earlier ones – and show how these principles emerged and how they work out in practice. We will also explore some of the problems, economic and other, associated with following them rigorously.

In part 2 we will present a varied body of empirical evidence concerning the effectiveness of professional development programmes. Most of this evidence has been reported previously only at conferences and here it will be laid out for closer inspection, and also collected together so that we can see how it accumulates and contributes to something like a unitary story. It comprises both quantitative evidence including gains in student achievement which can be attributed to the teachers' inservice courses, and also qualitative data obtained from questionnaires, interviews, and prolonged observations of classes of teachers participating in professional development (PD) courses.

There is, of course, already a substantial literature concerned with professional development in general, professional development of teachers, and the issue of educational change. We could not presume to offer new insights into effective professional development of teachers without recognising the enduring work of such scholars as Michael Fullan, Thomas Guskey, Andy Hargreaves, David

1

Hopkins, Bruce Joyce, Michael Huberman, Matthew Miles, and Virginia Richardson. But we have chosen to present our own experience and empirical data first and then, in Part 3, to show how this experience and data relates to models which have been proposed by others. We will address here methodological issues concerned with collecting and interpreting evidence of relationships amongst the many individual and situational factors associated with PD, and re-visit the arguments about 'process-product' research on PD. In the light of our experience, we will interrogate models of PD which have been proposed by others and attempt to move forward our total understanding of the process of the professional development of teachers for educational change. In conclusion, we will look at some current national practice in professional development, concentrating on the recent English experience of introducing 'strategies' into schools but referring also, by way of contrast, to the situation in the United States.

WHAT'S THE PROBLEM?

Why has the professional development of teachers already exercised so many good minds for so long? And how can we justify adding another book to this field? The answer to both questions must lie in the continuing demand from society in general (at least as interpreted by politicians and newspaper editors) for improvements in the quality of education. We are not here going to question the meaning of 'standards' in education, nor the validity of claims and publicly held perceptions about such standards, and we cannot be bothered to interrogate the motives of many of those who loudly express their horror at supposedly falling standards. (It is disappointing that even Michael Fullan (2001, p.47) occasionally makes glib statements such as "Most people would agree that the public school system is in a state of crisis") And we do not have to buy into the cataclysmic view of the rate of change in society fostered by futurologists such as Gleick (1999) and Toffler (1970) to accept that change is occurring, and that it is inevitable, is demanding of attention, and is welcome. It is welcome because a system which does not change is one which stagnates, and it demands attention, obviously, because methods of running a classroom, or school, or local authority, or government which worked fine 20 years ago will not work well now. The story of evolution is one of continuous change – sometimes so slow that we cannot detect it over hundreds of years, sometimes radically metamorphic. Species which fail to adapt become extinct.

Actually, the justification for continuing to chip away at the problem of professional development is quite simple. A desire to improve the quality of education is a perfectly respectable aim in its own right, and is one that will always continue, that should always continue, whatever successes may be achieved on the way. One school or one local education authority (LEA), or one country may achieve standards of instruction, provision of resources, and harmonious and productive relationships amongst teachers and students that would be the envy of the world, and yet still feel that more could be done, or at the very least that hard work must be put in to ensure that those standards are maintained. So, the raising and maintenance of educational standards is a continuous quest, and the central players in that quest must be the teachers.

> "Educational change depends on what teachers do and think – it's as simple and as complex as that" (Fullan & Stiegelbauer, 1991) p. 117

Hopkins & Lagerweij (1996) characterise decades of attempts to improve education as: 1960s, curriculum development and a belief that materials alone would do the job; 1970s, failure of this approach and much hand-wringing; 1980s, success of the school effectiveness movement in identifying key factors in successful schools, and the 1990s as the decade of pro-active school improvement. In his own 'Improving the Quality of Education for All' project, Hopkins noted as one of the important conditions which underpin improvement efforts a commitment to continuing staff development (p. 81). We would add, not just commitment, but an understanding of staff development methods which are effective.

Perhaps here an aside is in order about possible alternative routes to educational success which appear to by-pass teachers – the teacher-proof curriculum, the tightly specified lesson plan, and the computer-delivered lessons. It should not even be necessary to write this paragraph as we guess that it will be blindingly obvious to the great majority of our readers, but just in case someone[1] out there still believes in such by-passes, here goes. Education is first and foremost a social process, one that occurs between people. Whatever the de-schoolers or futurologists might argue, it is not an historical accident, nor a throw-back to medieval practice, nor the hopeless inertia of the system that has led all school education, everywhere in the world, to be conducted in 'classes' of from 15 to 90 students with one 'teacher'. The reason that the process of teaching and learning – even rather bad, didactic, teaching – can never be adequately replicated by a teaching machine, a computer, interactive video, or hypermedia text is that no machine can get near to managing the billions of subtle interactions which occur amongst even 30 students and between them and their teacher. Schön (1987) describes the *artistry* of the professional (including teachers) and the impossibility of reducing this artistry to some form of technical rationality. This is more than just saying that there are too many variables for a machine to handle for if that were all it was, machine development would soon catch up. It is that too many of the variables are indeterminate and manifest themselves anew whenever the context changes. Only a human agent can even approximate the most appropriate responses required to achieve a particular instructional goal with a group of other human beings. Human teachers are and will remain at the centre of the educational system, and thus the continuing professional development of teachers remains the most important force in the quest for educational improvement.

---

[1] At a dinner party while writing this book, the city types around the table agreed unanimously that star presenters of history and nature programmes on television were so excellent that they should be employed to make programmes to cover the school curriculum, which could then be shown in place of the teachers' 'boring' lessons. Then all the schools would need would be someone to take the register, turn on the video player, and stop fights in the playground. Obvious, really.

That is the justification for continuing to pay attention to professional development. The justification for this particular book will become apparent. It lies in our unique and prolonged experiences with extensive professional development programmes and the lessons which may be learned from them.

THE CONTEXT OF OUR WORK, AND ITS GENERALISABILITY

We would like to contextualise our work at two levels: within the school improvement literature in general and then as a specific example of professional development.

Professional Development and School Improvement

The radical conservative agenda of the 1970s in the United States and the United Kingdom turned an 'accountability' spotlight on to education. Education, it was argued, had much in common with any service industry: it had aims, outcomes, clients, stakeholders, and people who paid for it (generally the taxpayer). Since a main item of the agenda was to reduce taxes, the education system was required to account for its performance and to demonstrate cost-effectiveness. Notwithstanding a major problem in agreeing what counted as useful outcomes – for the manufacturer, these may be basic literacy and technical skills, for the service provider, interpersonal abilities and for the university admissions tutor, academic excellence – the political demand released funding for a wave of studies of educational effectiveness and educational change. This work has had a long-lasting impact both on methods of assessing effectiveness (e.g. Miles & Huberman, 1984) and on the construction of macro-models of educational change (e.g. Fullan, 1982; Fullan & Stiegelbauer, 1991). Early sociologists' suggestions that schools actually made very little difference to students' academic and personal development were countered by sophisticated longitudinal studies incorporating multi-level modelling which demonstrated unequivocally that school variables under the control of the school's managers did indeed have a significant effect on the variance in outcomes for students (e.g. Mortimore, Sammons, Ecob, Stoll, & Lewis, 1988; Rutter, 1980). Once it was established that schools did make a difference, attention could be turned to just how less-good schools could be improved, and the school improvement literature was born. Two excellent examples of this genre are provide by Joyce, Calhoun, & Hopkins (1999) and Stoll & Fink (1996), each of whom draws on experience, research, and imaginative analyses to present practicable ideas for improving schools by multi-pronged attention to a diverse range of parameters.

Although it constitutes only one of these prongs, the professional development of teachers is central to all plans for school improvement. In this book we will be recognising this centrality and focussing particularly on methods of effective professional development and its evaluation, but we recognise that this does only represent one of the aspects of the whole school situation which requires attention for substantial and lasting improvement. In short, professional development of teachers lies nested within school improvement, which in turn is part of the larger picture of educational change. Of course, the term professional development has a much more general applicability than to teachers only. Schön (1987), who is frequently quoted by educators for his wisdom on the idea of the reflective practitioner, was actually more generally interested in the university education of professionals, including architects and engineers as well as teachers. The department within which I work at King's College London is called the Department for Education and Professional Studies, a title much wrangled over but agreed because of the attention we pay to the professional development of health workers and priests as well as of teachers. The location of the subject of this book can thus be simply represented as a Venn diagram (figure 1.1).

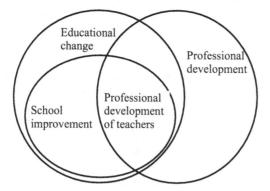

Figure 1.1: Locating the Professional Development of Teachers

By choosing a focus which is more limited than that of school improvement or professional development in general, we recognise that it is incumbent on us to offer a deeper treatment of the specifics of the professional development of teachers than might be expected from those wider literatures.

Cognitive Acceleration as a Context for Professional Development

The main work we are presenting and discussing in this book is a series of professional development programmes that we have run since 1991 to enable teachers to help their students develop higher level thinking. These are the "cognitive acceleration" programmes. There were important precursors to this work which we will describe in Chapter 2, but it is the professional development for cognitive acceleration which has set the greatest challenges and from which we have learned so much.

There are occasions where PD legitimately has quite limited aims, such as the introduction of a new assessment technique, or of a new textbook, or a new piece of software. Such aims may well be achievable by courses of one day or less because no attempt is being made to alter teachers' implicit models of teaching and learning and the aims of the inservice do not include shifts in beliefs and attitudes. They are limited to the introduction and practice of some straightforward technique. This is not the type of professional development with which this book is concerned. We are concerned with deep-seated changes in pedagogic practice which cannot be brought about without addressing both individuals' fundamental attitudes to teaching and learning and also a whole school's commitment to change. Teaching for the development of thinking does require a radical shift in pedagogy for most teachers, as well as fostering the support of colleagues within an institution. Cognitive acceleration also has a long and respectable track record of showing long-term effects on students' cognitive growth and academic achievement. It is thus an excellent example of an innovation which is of 'good quality' in the sense used by Fullan (1999, p. 80) and which makes heavy demands on professional development. Cognitive acceleration is a 'hard case': if PD for cognitive acceleration can be shown to be effective, then it must be doing quite a lot right and many of the same methods of the PD may be abstracted and applied to any professional development programme which aims at changing more than simple technical capability. We claim, therefore, that the lessons we have to offer on PD can be generalised far beyond the context of teaching for higher level thinking. (It should not, however, be thought that cognitive acceleration is a narrowly focussed context. In Chapter 3 we will say something of the international and inter-domain nature of cognitive acceleration in the 21st century.)

The point here is not that cognitive acceleration is being proposed as a 'magic bullet' or 'cure-all' system for raising achievement, although we think it's pretty good stuff. It is that it is the sort of innovation which requires deep level change in individuals and in institutions, and has lessons to offer far beyond that of a particular innovation.

A NOTE ABOUT AUTHORSHIP, AND ACKNOWLEDGEMENTS.

Most of the text of this book has been written by the first author, Philip Adey. But one of the main justifications for writing the book lies in the empirical data of Part 2 which has substantial contributions from Nicki Landau (especially chapter 8) and from Gwen and John Hewitt (especially chapter 9) and these colleagues are recognised as assistant authors. 'I' generally means Philip Adey, describing personal experiences and opinions, although in chapter 8 the 'I' is Nicki Landau. The use of 'we' generally demonstrates collective responsibility we have taken in drafting, editing, and refining the text although in chapter 9 the 'we' is John and Gwen Hewitt.

There are many others whose ideas and support have been invaluable over the years of experience of PD programmes, people we have worked alongside and from whom we have learned. A.J. Mee was a wise mentor in my early days of PD in the Caribbean. Benny Suprapto created the conditions for, and Gordon Aylward and Theresia Pietersz executed, one of the most remarkable PD programmes that has ever been implemented, to be described in chapter 2. I was fortunate indeed to be in Indonesia at the time and to have the chance to work with them. Through the years of developing PD programmes for cognitive acceleration I have learned much from watching, listening to, and being criticised by many colleagues and friends, including Carolyn Yates, Chris Harrison and Tony Hamaker, Anne Robertson and Grady Venville, Natasha Serret and Justin Dillon, and many others. We have also learned an enormous amount from the teachers we have worked with. Both explicitly and inadvertently they have steered the development of our methods and the development of our understanding of the process of the professional development of teachers. I am grateful also to the anonymous reviewer of the first draft of this book for his or her perceptive comments.

Finally, all of the cognitive acceleration work rests on the original vision of my mentor and friend, Michael Shayer with whom I have worked since 1974. The development of the theoretical models which underlie cognitive acceleration, and the generation and testing of hypotheses and of practical procedures have been developed through a constant process of dialogue between us in which I can always rely on Michael for deep insight and for connection to a remarkably wide range of literature and experience. Not least importantly, virtually all of the statistical procedures used for assessing gains by CA students were developed by Michael, and he was actually responsible for many of the specific analyses reported here.

2. EVOLVING PRINCIPLES: EXPERIENCE OF TWO LARGE SCALE PROGRAMMES

In 1970 I was not yet 30 years old, but after 7 years of teaching chemistry in a selective boys' school in Barbados I landed a job which seemed to me then (and still does now) to be like the best kind of dream. The job was to be the "Regional Consultant" in developing and introducing a new integrated science curriculum into junior secondary schools in 15 countries, almost all in the Eastern Caribbean. Mainly this was a curriculum development project to build on material (the West Indian Science Curriculum Innovation Project, WISCIP) which had been initiated by Iolo Wynn Williams and Harrod Thompson in Trinidad which in turn owed much to the then new Scottish integrated science curriculum. The relevance of this work to the present volume lay in the professional development programme associated with it. At this remove I am not sure how we designed the programme (or maybe I was never privy to that), but two important principles seemed to be taken for granted: (1) educational practice in schools would never be influenced by printed materials and kits of apparatus alone, however detailed the teachers' guide, however comprehensive the resource kit; and (2) even the addition of a series of 3-day inservice courses would be unlikely to have much impact unless we ourselves got into the classrooms to help the teachers implement the curriculum in their own context.

All of this was long before the ideas of 'coaching' or of 'school-focussed INSET' had been formalised, and yet it seemed obvious to us at the time. Nothing I have done in the field of professional development since then has disabused me of this belief. Rather, a further 30 years of experience (as well as much reading of others' experiences) has confirmed this simple truth: if you want to change what happens in schools, then you need to get into schools. For the first three years of the 1970s, virtually every term-time week saw me on a little plane to St. Vincent, to St. Lucia, or Dominica, occasionally further afield to somewhere like the Cayman Islands or Turks and Caicos, driving around the islands with school inspectors, observing lessons and offering feedback, listening to teachers' stories of success and difficulties, and learning all of the time from their experiences.

Although the project did put good resources into about one hundred schools, and did provide the teachers with some sort of entrée into constructivist science teaching, it was not an unmitigated success for all of the schools involved. The limited nature of my own experience, and even the limited nature of my own subject knowledge outside of chemistry, led me to make mistakes and some crass suggestions. It took time for it to dawn on me that conceptual teaching which had worked pretty well with my bright grammar school boys could not simply be transferred unaltered to a class in an all-age school, from which the grammar stream children had been selected out, in up-country Guyana. (That this realisation eventually led me back to university to research cognitive development may have been a useful pay-off for me, but it didn't do a lot at the time for the children or their teachers.)

At this remove I would hesitate to comment on the cost-effectiveness of the WISCIP project. That was a time when overseas development agencies of British and American governments had great faith in science education as a long-term route to economic development and WISCIP was probably a lot better than many projects which dropped kits of apparatus into schools around the world, enriching equipment manufacturers but leaving many an under-educated science teacher bemused at boxes which had arrived without any prior consultation about what might actually be useful.

In retrospect I would say two strong messages started to formulate themselves from the WISCIP experience about effective professional development, in addition to the faith in in-school coaching and the realisation that there were no quick fixes. The first arose from my observation that, even after two years, I would so often "just miss" seeing a science lesson in a school which I visited. The time and date of my visit would have been sent in advance, but somehow the lesson was going to be tomorrow, or had been yesterday, or the children had been called to a sports day, or the keys couldn't be found to the lab (yes, really). Changing your teaching practice is a frightening thing. Once you move from dictating notes from the textbook (possibly the only experience you have had of education thus far) you are entering an uncertain world. You don't want to do it in front of a stranger. And you don't want to do it by yourself. The safest strategy is to lose the lab keys. What this says about coaching is that coaching is not dropping in off an aeroplane to observe a lesson, make some encouraging comments, and moving on the next island. We will have more to say about what coaching *is* in chapters 4 and 11.

The second lesson is more positive. It is that the best experience that teachers have in inservice courses is talking with other teachers. In the WISCIP context teachers would fly in from all over the Caribbean to a campus of the University of the West Indies in Barbados or Jamaica, stay in student accommodation for a few days, take part in WISCIP workshops during the day – with plenty of opportunity for participant interaction – and then socialise in the evening. They learnt far more from each other than they ever did from us. The teacher with a bullying headteacher learned that she was not alone. The man who had said WISCIP was impossible without a fully equipped laboratory learned how others were coping. Most of all (and this only after a few sessions) teachers learned that others were as vulnerable as they were, and that help and sympathy could be obtained by sharing fears, difficulties, and 'errors'. The process of change became a little less frightening.

THE INSERVICE-ONSERVICE MODEL IN INDONESIA[1]

The Dutch colonial powers in the East Indies did not set much store on the education of the indigenous people, and when Indonesia finally gained independence in 1948 (after a 'police action' in which the British helped the Dutch try to regain their former territories after the defeat of the Japanese in 1945) it was faced with establishing an education system virtually from nothing. For Sukarno, leader of the independence movement who became the first President of the Republic of Indonesia, the immediate priority was to weld together as one country a set of five large land masses and some 3000 smaller islands, covering an area about 5000 km East to West by 3000 km North to South and population of around 150 million incorporating a bewildering diversity of languages, religions, cultures, and levels of development. From his power base in Java he did this remarkably successfully through a combination of clever diplomatic moves (such as making a variant of the widely used Malay trading language, Bahasa Indonesia, the national language rather than try to impose Javanese, and by consulting closely with the Governors of the 27 Provinces) and force, using the predominantly Javanese-led army which had successfully harried the Japanese and made life so miserable for the Dutch and British that they had to abandon their 'police action'. In line with this political welding together of the country, the education system was massively centralised. General school education had three levels: elementary (*Sekolah Dasar*, SD), junior secondary (*Sekolah Menengah Pertama*, SMP) and senior secondary (*Sekolah Menengah Atas*, SMA). The system grew at an extraordinary rate, such that by 1970 the percentages of the population able to attend each level of this system were something like 80% at SD, 40% at SMP, and 15% at SMA. Teacher Universities (*Institut Keguruan dan Ilmu Pendidikan*, IKIP) and faculties of education within regular universities (FKIP) were established but their capacities (and capabilities) fell far short of the demands of the rapidly expanding system. Thus the achievement of educational expansion came at the cost of years of emergency training of teachers, some of whom received no more than one year teacher training beyond the level at which they were expected to teach. Even those fortunate enough to complete a 2 or 3 year Diploma, or even a 4 year degree, at an IKIP or FKIP generally experienced a programme in which subject matter content and educational theory were rigorously separated and taught by different departments (Van den Berg, 1984). University-level education departments in Indonesia suffered the same problem as education departments in many parts of the world (notably the United States): in fighting to establish their academic credibility in a university environment, they so fear being labelled 'vocational' that they lose touch with the reality of schools and become overly academic. This syndrome has been well documented by Donald Schön (1987).

---

[1] Much of the information of this section comes from personal experience of working as a British Council Education Officer, seconded part-time to the Indonesian Ministry of Education and Culture, from 1981 – 1984 and many subsequent visits as a consultant on Governement of Indonesia, World Bank, or UNESCO teams.

The quality of education in schools, inevitably, reflected this under-educated teaching force, and the problem was exacerbated by the low pay of teachers which meant that they generally had to teach two shifts - the same building being used for different schools in the morning and in the afternoon - in order to make a living. There is no time for preparation or marking in this situation.

From the 1970s there was a concerted effort to address the problem of quality by running national inservice workshops for selected teachers and local inspectors. The focus of these workshops was mostly child-centred learning and activity-based teaching methods, although some attention was paid to teachers' content knowledge. Participants came from all of the 27 provinces, and on their return were supposed to cascade their new knowledge through a series of steps in order to reach as many teachers as possible. From the perspective of the 21st century and all that has been written about cascade methods, it is hardly surprising that this effort, although well-meant and rather expensive, had very little impact on practice in schools.

In 1979 the Department of Education and Culture of the Government of Indonesia initiated a project called PKG (*Pemantapan Kerja Guru* - improving the work of the teachers) which was to become the biggest inservice project anywhere in the world. Funding was obtained initially from UNESCO and UNDP, and later the World Bank became involved. Since its inception PKG has had an ambitious objective: to learn from previous attempts at cascade models of inservice education and to reach directly into thousands of junior and senior secondary schools (SMP, SMA) throughout the 27 provinces of Indonesia. The aim was to shift the locus of control, if only slightly, from teachers as purveyors of knowledge (or, commonly, as mediators of the knowledge enshrined in a textbook) towards the students as active constructors of their own understandings.

From the start, the originators of the project – Dr. Benny Suprapto, Director of Secondary General Education; Theresia Pietersz, National Project Consultant, and Dr. Gordon Aylward, a consultant from Australia – were committed to the idea of a substantial onservice element – that is, work with teachers in their own schools as well as at inservice days organised in central locations. As with the assumption we had made in the Caribbean, this decision was probably based less on the theoretical considerations than on the originators' combined experience of inservice teacher education projects in various parts of the world. Expensive but centrally-based projects had too often been seen to founder as soon as external funding terminated. Another lesson that had been learnt was that although the IKIPs and FKIPs contained some talented individuals, as institutions they were more likely to be obstructive than facilitating in the process of pedagogical change, for the reasons outlined above. It was therefore necessary to work around them, which in the Indonesian context required some deft political foot-work.

An immediate question is, how can one possibly provide in-class coaching to many thousands of teachers spread over such a vast and varied area where communication in the many remote regions was often extremely difficult? What would be the cost and logistics involved in deploying the army of teacher-coaches required? PKG's solution was entirely appropriate to the situation: draw coaches from the ranks of teachers themselves. Over the years there have been many modifications to the detail, but the basic model involved the following steps:

1 With the help of provincial and district authorities, use criteria including experience, qualifications, and tests of content knowledge to select a cadre of existing teachers from the secondary schools who seemed to be better qualified and motivated than the average.

2 Take this cadre of teachers for specialist training workshops at provincial, national, regional, and on occasion at international workshops.

3 Return them to their own schools for a semester, where they practice for themselves the methods they have learned, and meet weekly in small groups to share experiences. National consultants make inputs and monitor these meetings.

4 Accredit this cadre of trained and experienced teachers as 'Instructors' or Assistant Instructors (depending on the training they had received). Later another layer of coach was recruited, the Key Teachers (*Guru Inti*). Instructors now become responsible for running the main training programmes in their Provinces. These programmes last one semester each, and consist of :

   a.   a two week introductory workshop
   b.   meetings every Saturday throughout the semester
   c.   coaching visits by the trainers to teachers in their own schools during the week.
   There are also a mid-semester one week workshop and an end-of-semester one week workshop for reflection and transfer work.

5 Trainers and key teachers meet at annual national workshops to reflect on their coaching practice and to develop new content materials.

There are many more aspects to this vast professional development programme than can be detailed here, but in summary it operated as a sort of two or three-step cascade, but with a critical added element of feedback up the cascade, and continual monitoring by the national team of national and provincial workshops to maintain quality and guard against the classic dilution effects which beset standard cascade models. The feedback process ran in a series of loops from teachers' own inputs to the development of materials and their evaluation of materials provided, up through local, provincial, regional, and national training and writing workshops. There was a remarkable degree of consultation at every level. (For a more detailed account of the PKG system, see Thair & Treagust (2003), although we do not necessarily buy into the implications which they draw.)

Evaluation

When I joined the project in a part-time capacity in 1981, it took me perhaps six months to move from scepticism that such a grandly conceived programme with such a small central team could possibly have any effect at all, through a phase of incredulous wonder, to one of more evidence-based conviction that here was a model which, at least in the highly consensual culture which characterises most of

Indonesia, was at the time enormously powerful and cost-effective. Over three years I observed schools and local PKG workshops from Aceh (the Northernmost province of Sumatra with a strongly traditionalist Islamic culture) through North Sulawesi (predominantly Christian) to Iryan Jaya (the Western half of New Guinea, where most settlements are accessible only by light aeroplane and the people largely retain animist beliefs). Certainly I saw a lot of bad teaching and bad instructing, but overall the teachers' ability (and willingness) to engage with their students and promote active learning methods was quite remarkable both in scale and in the light of the previous paucity of educational background of the majority of the participants. It is probable that the feedback loop system worked so well for PKG because it harmonised with a culture which values consensus above the rule of the majority and is prepared to devote many hours in discussion in order to reach such consensus.

An early, more formal, evaluation of PKG by Egglestone (1984) was based on classroom observations and assessments of student group[1] practical work in matched PKG and non-PKG classrooms, supplemented by collecting grades on nationally set examinations of content knowledge. He reported statistically significant positive effects in the PKG classes on student active participation in the learning process, on their practical science problem solving ability, and small but non-significant gains in the national science subject matter assessments. The last of these is important since teachers and administrators often expressed the fear that the extra time spent in PKG classes on practical work and constructive discussion might adversely affect the students' scores in the national tests. That students maintained expected levels of recall knowledge while experiencing a far richer programme of constructivist teaching was an important finding.

Thair & Treagust (1997) report a series of quantitative studies of the effects of PKG-style teaching on student learning, these studies being executed by PKG Instructors while doing Masters' degrees in science education at Curtin University (Western Australia). While these show clearly that PKG style teaching is beneficial to students' understanding and knowledge development – and thus confirm that PKG is a 'good quality' innovation (Fullan, 1999 p. 80) - they do not point unequivocally to the success of the PKG inservice-onservice project itself since in some cases both experimental and control groups were taught by PKG trained teachers. There is nevertheless clear evidence from these studies that those teachers who take on the PKG message do improve the quality of their teaching and the achievement of their students. My own observations - extensive if less systematic - indicated that it was a clear majority of those who participated in PKG who made real changes in their practice.

Unfortunately this is not the simple happy end to the story. More recent diagnostic surveys (Blazely, Samnai, Rahayu, & Purwati, 1996; Sadtono,

---

[1] In designing the evaluation, Jim Egglestone recognised that all practical work was conducted in groups of about six students, and that assessing individual practical skills would be culturally inappropriate. His team thus developed practical problems to be solved by groups.

Handayani, & O'Reilly, 1996; Somerset, 1996) and an investigation assisted by Tony Somerset (Mahady, Wardani, Irianto, Somerset, & Nielson, 1996) have all indicated that those initial gains have not been maintained. Although none of these diagnostic surveys was able to correlate data on teachers' inservice experience with the classroom observations, the overall impression is given that the early effects of PKG on the quality of teaching have not been maintained. Mahady, et al. (1996) suggest three possible reasons for this:

1   the growth in size of the programme, leading to a dilution of the influence of the central team;
2   the loss from the programme of the onservice visits to provide in-class support to teachers trying new methods;
3   the addition of an extra step in the 'cascade' from National level to classroom level.

Of these, (1) may be the weakest hypothesis since PKG was already large, operating in 27 Provinces, in 1984. (3) may be a factor but it is one which is specifically addressed in the most recent form of the project where a two-way trickle-down, and feedback-up loop maintains the continuing development and monitoring of trainers at every level in the system. It seems to be most likely that it is (2), the loss of onservice visits, which must bear the main responsibility for the loss of effectiveness of PKG. The evidence here, that an immense programme which was successful as long as the onservice (in-school) work was maintained, appears to go into decline when that onservice is curtailed, may be seen as just another nail in the coffin of inservice programmes which make no provision for in-school coaching. This coffin was ably constructed by Joyce & Showers (1988) in the first edition of 'Student Achievement through Staff Development', where they report from a meta-analysis of effective staff development that professional development without some form of in-school coaching is, at best, a waste of money and effort.

LESSONS TO TAKE FROM THE CARIBBEAN AND FROM INDONESIA

Even without the benefit of Joyce & Showers' (1980) meta-analysis of coaching, or the OECD work on school-focussed INSET (Hopkins, 1986), or the comprehensive synthesis of research on effective professional development of Fullan (1982), when I returned to England in 1984 to start work with Michael Shayer on cognitive acceleration, the experience of two large-scale PD projects had seemed to establish some ground rules beyond reasonable doubt. Change in schools, change in pedagogy, demands attention to *at least* the following principles:

1   There is no such thing as a teacher-proof curriculum. Whether one takes a narrow view of curriculum as a set of planned and written-out teaching activities, or a broader Stenhousian (Stenhouse, 1975) view of curriculum as all of the interactions between children and their teachers with are directed towards learning, the process of curriculum 'development' implies a change in teaching methods. That cannot be brought about in any meaningful way except by

working directly with teachers, by providing some form of professional development which is much more than 'showing them how to use the materials'.

2   Change cannot be imposed. Teachers must be brought into the process of change as partners. This does not mean that whatever teachers say they want is what they should be offered, since programme initiators will by definition have a clearer vision of what the programme is about, the justification for its aims and methods and where, roughly speaking, it is headed. They have a responsibility to lay out what the possibilities are and to provide information on research evidence and philosophical positions. Teachers are partners nevertheless, who are genuinely consulted and listened to, and who oftentimes provide real learning experiences for the project leaders.

3   In-class coaching is essential. Much has now been written on coaching, and we will review some of this literature in chapter 11. For now, we need note only that coaching can take many forms, including demonstration lessons, classic observation-plus-feedback, team teaching, peer-coaching, and video-based feedback. Whatever its format, it plays the critical role of bringing the practicalities of pedagogical change into the teachers' own classroom with their own students.

4   Change is slow, uncertain, and has many backward steps as well as forward ones.

Over the next 20 years many more principles of PD accrued to this basic structure through further experience and reflection, enhanced by specific empirical research and reviews of the experience and research of others. In the next two chapters we will show how the principles develop through intensive experience.

3. PROFESSIONAL DEVELOPMENT FOR COGNITIVE ACCELERATION: INITIATION

OVERVIEW OF COGNITIVE ACCELERATION

Around 1981, after a decade of investigating and assessing the cognitive development of the school population of England and Wales, Michael Shayer turned his attention to what Piaget had called "The American Question": can cognitive development be accelerated? The question begs all sorts of further questions such as "Accelerated relative to what?", "Do children have some maximum potential which is not normally reached?" and "What counts as 'normal'?", but a book devoted to professional development is not the place to go into them. We have essayed answers in *Really Raising Standards* (Adey & Shayer, 1994). Here it is sufficient to characterise 'cognitive acceleration' as an intervention in children's education designed to promote, to enhance, their progress through the process of cognitive development so comprehensively charted by Jean Piaget and his co-workers at the University of Geneva.

By 1983 Shayer had had enough encouragement from effects with one class in a Sussex school to submit and to have accepted a major proposal to the Economic and Social Research Council of England for a full scale trial of Cognitive Acceleration through Science Education (CASE). I returned from Indonesia to become senior researcher on the project and we recruited Carolyn Yates as the third member of the team. We started work in September 1984.

If we look at cognitive acceleration today we see a wide range of programmes operating with many age groups, in the context of many different subject areas, with trials in many countries across the world. Table 3.1 summarises these programmes in the UK and Table 3.2 indicates some of the spread of CA internationally.

Table 3.2: Cognitive Acceleration programmes in the UK, January 2003

Project	Age range	R&D Phase	Funding	PD Phase	Published materials	Added-Value Assessment	Main researchers
CASE	12-14 yr.	1984-1987	SSRC	1991 onwards	*Thinking science*, Nelson 1989	GCSE 1995 on and KS3 tests	Shayer, Adey, Yates
CAME	12-14 yr.	1993-1997	Leverhulme E. Fairbairn, ESRC	1997 onwards	*Thinking Mathematics* Heinemann 1998	GCSE 2001 on and KS3 tests	Shayer, Adhami, Johnson
CATE	12-16 yr.	1994-2000	Greenwich LEA?	2001 onwards	*CATE* Nigel Blagg 2002	KS3 2001	Hamaker, Backwell
Wigan ARTS	12-14 yr.	1999-2002	Wigan LEA		Wigan LEA	KS3 2002	Gouge, Yates, Wigan ARTS grpou
Thinking Arts	9 – 11 yrs	2002-2003	Cognitive Acceleration Programmes	2003 onwards			Gouge and Yates
CAME@KS2	9-11 yr.	1997-2000	Leverhulme Trust	2001 onwards	BEAM 2002	KS2 2003 onwards	Johnson, Adhami, Shayer, Hafeez
CA@KS1	5-6 yr.	1998-2001	Hammer-smith LEA	2001 onwards	*Let's Think!* NferNelson 2001	KS1 2001 and 2002	Adey, Robertson, Venville
CASE@KS2	7-8 yr.	2000-2002	Astra Zeneca science	2002 onwards	*Let's Think Through science* nferNelson 2003		Wilson, Adey, Dillon, Robertson
CAME @KS1	6-7 yr.	2001 -	ESRC				Shayer, Adhami, Robertson

Table 3.2 Some international applications of Cognitive Acceleration

Australia: Perth	Grady Venville, Curtin University, is introducing *Let's Think!* into one school and planning further research
Australia: Townsville	Lorna Endler & Trevor Bond (2001) implemented CASE with a cohort of three grade 8 classes and report extra cognitive growth between grades 8 and 10 for most students across the ability range with significant correlation between cognitive development and the scholastic achievement
Finland	Jarkko Hautamäki, Helsinki University, has been conducting cognitive assessments since the 1970s and introducing cognitive acceleration since the '80s. Recently, with Jorma Kuusela, they have conducted an extraordinary experimental test of the effects of CASE and CAME in a town's school system (Hautamäki, Kuusela, & Wikström, 2002)
Germany	Adey, Shayer, & Yates (1993) is the German version of *Thinking Science*. There were some trials of the material in schools associated with the University of Bremen.
Holland	Martin van Os and Peter van Aalten have introduced *Thinking Science (Denklessen)* into many schools in Holland, where it is taught by non-science teachers in a social period.
Israel	We believe that there has been some unauthorised translation of *Thinking Science* into Hebrew.
Korea	Byung-Soon Choi at the National Teachers University and Jeong-Hee Nam at Pusan University have made extensive trials of CASE (Choi & Han, 2002; Nam, Choi, Lee, & Choi, 2002).
Palestine	Around 1997 Carolyn Yates worked with the Palestinian Education Authority to introduce CASE into West Bank schools. The current destruction of the education authority in Ramallah has halted further progress on this, but Dua' Dajani of Qattan Centre for Educational Research is translating and introducing *Let's Think!* into some primary schools in Gaza.
Slovenia	Dusan Krnel of the University of Lublijana is leading a group of CASE trainers to introduce CASE into schools.
USA: Arizona	About 1992 the Glendale school district in Phoenix introduced CASE into all of its high schools and also wrote many more activities so that cognitive acceleration became the science course of the freshman year. Jim Forsman was the District science inspector responsible, and Jolene Henrickson (née Barber) a teacher who devised many of the new activities (Forsman, Adey, & Barber, 1993).

USA: Oregon   A Scientific Thinking Enhancement Project was set up in 1999
               and student progress was monitored by Endler and Bond (2001).
               The intervention was implemented with three cohorts of in 6th,
               7th and 8th grades. Preliminary results show gains in cognitive
               development for STEP students in all three cohorts.

Tables 3.1 and 3.2 offer snapshots of the breadth and depth of Cognitive
Acceleration (CA) as we know it in 2003, some 20 years after Shayer's initial
exploration, but the growth to this point has been very uneven. The original CASE
project was an intervention designed to be delivered over two years to students aged
12 – 14 (Years 7 and 8, the start of secondary education in England; grades 6 and 7,
middle school in many parts of the USA). The intervention consists of a special
activity to replace a science lesson once every two weeks. That this activity makes
a heavy demand on the teacher's skill and understanding is, of course, the main
driver of the professional development programme associated with CASE and so is
the mainspring for this book. For the time being, we need only note these key
milestones in the development of cognitive acceleration over 20 years:
1984: Start of CASE II project with Shayer, Adey, and Yates writing and trialling
       activities in two London schools
1985-87: Trial of the materials, pedagogy, and PD in 10 schools. Quasi-
       experimental design reveals significantly greater cognitive development and
       delayed effects on science achievement in experimental classes. Publication of
       initial results in academic science education journals (see chapter 5).
1989-91: 'CASE III': Michael Shayer explores scaling up the PD in 3 schools from
       one teacher to whole department.
1990: Long-term follow-up reveals that CASE has far transfer effects on students'
       achievement in national public examinations in mathematics and English as well
       as in science. National publicity for this effect in England leads to heavy demand
       from schools. Start of the first two-year PD programme for CASE.
1994: Establishment of CAME (CA in Mathematics Education).
       As will be seen from table 3.1, the other CA programmes came tumbling in
since 1999, when the initiation of a project for 5 and 6 year olds opened a
completely new age group to the possibility of CA and the work of Ken Gouge and
Carolyn Yates established the feasibility of CA in the visual, musical, and dramatic
arts subjects. More detail of this work is available in Shayer & Adey (2002). At the
time of writing, it feels like we have a solid base of CASE and CAME which
continue along fairly well-established paths, and a bunch of exciting possibilities
stemming from these foundations, some of which may peter out in the sand, some of
which will certainly grow strong and secure.
       We need now to look at what all expressions of CA have in common: their
underpinnings in cognitive psychology.

THE THEORETICAL BASES OF COGNITIVE ACCELERATION

Describing the theory which underlies CA is more than an academic exercise in a book devoted to professional development. To appreciate the requirements of the CA PD programme, one needs some insight into the aims and theoretical underpinnings of cognitive acceleration. Pedagogical methods for CA arise from its psychological theory base, and acquaintance with this theory base offers some understanding of the challenges facing the professional development programme. Here we will summarise the main features of that theory base and point to some features of the pedagogy which are implied by the theory.

There are three central 'pillars' to cognitive acceleration: cognitive conflict, social construction, and metacognition. The notion of cognitive conflict comes from the Piagetian principle of equilibration: when the mind encounters a problem which requires a somewhat more sophisticated cognitive structure than is currently available, it attempts to grow to meet the challenge, to accommodate to the new demand. Clearly the level of conflict must be no more than moderate, since if the demand is excessive the mind simply makes no sense of it at all. To tell the same story from a Vygotskyan perspective, this principle of cognitive acceleration requires that the student is working within their zone of proximal development – what Newman, Griffin, & Cole (1989) call 'the construction zone'. It is the intellectual zone which is just beyond the student's current unaided capability, where they are struggling somewhat, and where they need well structured, scaffolding, support. CA activities and teaching methods are designed to maximise the opportunities for cognitive conflict.

The implications of this 'pillar' for pedagogy should be clear, but why is it difficult for teachers to maintain this level of cognitive conflict? The answer lies partly in the fact that teachers are essentially nice people. They get very uncomfortable watching their charges struggling, and too often rush in with answers which they believe will be helpful but which, in the context of cognitive acceleration, actually short-circuit the process. Another reason that managing cognitive conflict effectively is so difficult is that the construction zone is going to be different for every child in the class. This may seem to a more intractable problem than the first, but the reality is that teachers soon learn how to manage this difficulty, while learning to live with their students in cognitive discomfort requires a more fundamental shift in their whole attitude to classroom processes.

The second 'pillar', social construction, calls directly on the best-known feature of Vygotskyan psychology, that learning and the development of intelligence is essentially a social process. This is far more than a matter of becoming socialised into a set of beliefs by the cultural milieu in which one finds oneself. What it means is that our ability to process information, the actual development of intelligence, depends critically on social interaction, on the chucking back and forth of ideas and challenges, defending a position and learning to give up an untenable position gracefully. Good CA lessons include a great deal of on-task discussion and constructive argument in small groups and between groups, with each individual or group learning (sometimes over many weeks) how to put their position, how to be tentative, how to listen, how to challenge politely, and how to take risks. Managing, let alone encouraging the development of this type of classroom discussion is not a trivial task. It does require an approach to the classroom and what goes on in it which is significantly different from what is normally considered to be good quality conceptual teaching. It is not a small thing to ask a teacher who has successfully mastered the craft of setting clear objectives for a lesson, and generally attaining them, to occasionally abandon the comfort of knowing where the lesson is going and value instead the quality of the argument, wherever it may lead.

Finally, metacognition: becoming conscious of one's own thinking is now well-accepted as a powerful strategy within many effective learning models (Brown, 1987; De Corte, 1990; Hennessy, 1999; Kuhn, 1999; White & Mitchell, 1994). We would again draw on Piaget, for his notion of reflective abstraction, which he saw as essentially a formal operation available to late adolescents for whom the discussion of possibilities (and comparisons with actualities) is so important. But unlike Piaget, we see metacognition as available 'in some intellectually honest way' to any child who has developed a theory of mind – that is, generally to children from about 4 years of age. The process of putting one's thoughts into words, of reflecting back on what I thought an hour ago, what I think now, and why I have changed, is also closely linked with Vygotsky's emphasis on the use of language as a mediator of thought. We suggest that metacognition actually plays two separate roles in the process of cognitive acceleration. There is the intellectual role, which involves both the challenge of verbalising thought (offering its own cognitive conflict) and the value of explicating thought so that the same thinking is more readily available for use on another occasion. But there is also an affective role, whereby the process of exposing one's thinking makes a student aware of the fact that he or she is a thinker, can solve problems, does know how to seek assistance of colleagues, and can overcome initial incomprehension. This is closely related to the process well described by Carol Dweck (1991) of shifting the students' notion of their own intelligence from something over which they have little control ("I'm just stupid", "I'm just clever") to something more 'incremental', something fluid which can be developed in a manner akin to the development of muscle by appropriate exercise.

Learning to manage cognitive conflict and social construction causes many teachers real headaches, but we have found that learning how to generate metacognition is the most difficult 'pillar' of all for teachers to manage. As a general rule, it is not until the second year of the CA PD programme that participating teachers are becoming adept at getting their students to probe their own thinking processes in a valuable way. I have to say I am not clear why this should be so. It may be that until teachers have learned to question their own beliefs, have found ways of reflecting on their own practice in an open and non-defensive manner, they find it difficult even to comprehend the nature of metacognition, let alone encourage its development in their students.

Of this three-pillar model, we have sometimes been asked which of the three we believe to be doing the real work. It turns out that this is impossible to answer, since it is difficult to imagine an experiment, even in the conditions of a psychological laboratory, where the three variables could be independently controlled. In classroom practice it would be simply impossible. Well-managed cognitive conflict (struggling with a problem) almost guarantees that students will talk with one another, will construct new understanding socially. And reflecting on the thinking process as it happens, or afterwards, again generates cognitive conflict and the process of putting thoughts into words in public is itself a process of social construction. We have to see the three pillars as three facets of a process which, at its best, is an integrated whole.

To these three central pillars we must add two more: 'concrete preparation' is the early phase of a CA lesson when the teacher introduces the context of the activity and any new words which will be encountered. This in itself is not demanding, but sets the scene in which the cognitive conflict can be engendered without the extra confusion of unfamiliar language and apparatus. Then, often near the end of the activity, there is a process of 'bridging', in which the thinking developed during the activity and explicated through metacognition is applied to different contexts: "Where else might we use this sort of thinking?" Bridging may be into other areas of science, into other subject areas, or into the normal world outside school.

Table 3.3: Comparison of CA-type intervention and normal good instructional teaching

Instruction	Intervention
Carefully ordered	Follow direction of argument
Specific objectives	Virtual objectives
Small packets, reinforced	Students often puzzled
Lots of stuff delivered	Not much stuff delivered
Students have notes to revise	Nothing obvious to show
You know what you have covered	Not sure what you have covered
Relatively easy	Seems dangerous

Those then are the characteristics which are common to all CA programmes – they are, if you like, the minimum criteria by which CA is distinguished from other educational programmes. The pedagogical implications which follow from the psychological model are summarised in table 3.3 which compare teaching strategies typical of cognitive intervention lessons with those of even high quality teaching aimed at the development of conceptual understanding. The point here is not that teaching for cognitive acceleration is always better than teaching for conceptual understanding, but that both styles of teaching are complementary to one another, each offering particular sorts of outcomes.

Notwithstanding the common elements across all CA programmes, within the family of CA there are some variations. Perhaps the most obvious is the subject matter – science, mathematics, the arts, or technology – which offers the context for the development of general intellectual processes. But a more important distinction relates to the age groups at which the CA programme is aimed, and this distinction again owes much to Piaget's description of the development of cognition. The original CASE and CAME programmes, designed for 12 – 14 year olds, took Inhelder & Piaget's (1958) characterisation of the schema of formal operations as the 'subject matter' of the intervention activities. While on the surface the activities look like science or mathematics, the deep structure of each activity is in fact one of the schemata of formal operational thinking such as proportionality, probability, equilibrium, or compensation. These formal schemata would be far too demanding for the youngest children engaged in CA so for the 5 years olds, the schemata of concrete operations such as simple classification, causality or seriation provide the 'subject matter' of the activities.

Since setting cognitive challenge at the right level requires some understanding of the nature of the schemata – their characteristics and how they become elaborated over the years of development – this too must be included in the professional development of teachers for cognitive acceleration.

For the remainder of this chapter and the bulk of the next chapter, we will concentrate on the professional development programme established for CASE – that is the science based cognitive acceleration project for secondary schools. The last section of chapter 4 will discuss differences the PD programme for primary schools, its similarities and differences from the CA PD for secondary teachers.

DEVELOPING THE PD PROGRAMME FOR CASE: 1984-91

In 1985 when we tried out the first version of CASE in 10 schools our main preoccupation was with the evaluation and development of the activities and the testing programme to assess the effect of the programme on the students. We knew well enough from our previous experience (see for example chapter 2) that implementation of curriculum activities necessarily entailed the professional development of the teachers, including in-school coaching, but at this remove I do not remember that we ever sat down as a team and planned out a programme of so many days of inservice with so many school visits, with the content of the programme mapped out in advance. In those two years it was much more a matter of working closely with the teachers, eliciting from them of the sort of support they wanted, getting a sense from our time in classes of the features of CA which needed special attention, and developing the inservice teacher education programme as we went along. To be sure, we had a lot of experience-based intuitive knowledge about the process.

There is one lesson from this phase of the professional development which fed usefully, if negatively, into the main model of PD that we subsequently developed: an individual teacher finds it virtually impossible to maintain a radically new form of teaching while colleagues around them in the same school remain untouched by the innovation. During the phase of materials development and evaluation, CASE was tried out with only one teacher in each of 10 schools. Other classes in the same schools acted as controls. This meant that within each school, the teacher trying out CASE activities for the first time really had no one with whom she or he could discuss what was happening, at least no one with the experience themselves of teaching CASE. The fact that the great majority of the trial schools did not continue with the programme for long after the end of that experimental project must partly be attributable to this absence of a critical mass of teachers with the appropriate experience, although the introduction at that time of a national curriculum in England would also have created an obstacle for a school wishing to maintain the CASE programme.

We have touched already (in Chapter 2, on WISCIP) on the value of teachers sharing experiences with one another. This is a teacher-level manifestation of one of the pillars of CA: knowledge and understanding (and skill) is constructed socially. At this level it means that for change to take root in a school it is essential for the teachers involved to be able to work together in sharing and discussing the new methods with one another as the change is gradually implemented. The very least that is needed is for the teachers involved to be able to talk informally about the successes and difficulties they are encountering, and to do this in a mutually supportive atmosphere where problems and apparent failures are seen as learning opportunities.

This was one of the purposes of the 'CASE III' project undertaken by Michael Shayer from 1989 to 1991. He worked closely in three schools to scale up the introduction of CASE from one class to the whole science department. Although the main aim of this work was to gain deeper insight into the classroom processes which maximise cognitive stimulation, a valuable by-product was the experience of working with a number of teachers together in each school, providing the catalyst which overcame their activation energy barrier and enabling them to share their experiences with one another.

THE PD PROGRAMME FOR CASE: 1991 ONWARDS

By September 1990 we had data from the original CASE experiment which showed that students who experienced the programme when they were in Years 7 and 8 went on to score significantly higher grades at GCSE[1]. This was published in academic journals (Adey & Shayer, 1993; Shayer & Adey, 1993) but was also reported in the national press and was the subject of a television documentary. By May 1991 the long-term effects of CASE on academic achievement – in a political environment which emphasised 'raising standards' in education – had become sufficiently widely publicised to attract considerable interest from headteachers and science teachers who believed that CASE would provide a valuable addition to their armoury of strategies for raising academic achievement (and for the continuing professional development of their teachers). The title of our 1994 book, *Really*

---

[1] GCSE - General Certificate of Secondary Education – is a nationally set and marked examination taken in each subject by students at the end of Year 11, when they are about 16 years old. GCSE is graded A* down to G, plus 'U' for unclassified. Schools' GCSE grades in each subject are published each year.

Raising Standards, testifies to our scepticism about the then Government's notions of what an educational 'standard' might actually look like, and we did then and do now take a somewhat more sophisticated view of the purpose of education than the achievement of high GCSE grades. Nevertheless, GCSE grades are a currency in which a school's educational value is measured and the fact that CA seemed to improve GCSE grades, whatever other more important cognitive gains might be involved, inevitably attracted attention in the educational media and in schools.

There was an exciting and somewhat frightening few weeks for us in May and June 1991. The demand for a programme to introduce CASE to schools had suddenly been stimulated, but we had no ready-made programme to offer. We answered the enquiries from schools with great confidence while at the same time rather desperately sketching out a possible PD programme. At the time this seemed like something of a shot in the dark, although in retrospect it is easy to see that the programme did in fact have very sound foundations in our by now considerable experience, as well as in the experience of others and in the academic research on PD which was becoming substantial by this time.

It seemed clear that the structure of an effective PD programme for CA must include at least the following features:
- it must last for at least two years, paralleling the two year CASE programme itself;
- it must include both centre-based inservice days and in-school coaching;
- it must involve all members of the science department in a school;
- the inservice programme needed to be front-loaded, with the majority of centre based days near the beginning of the two years but with contact continuing throughout the PD period (and if possible beyond).

As to the content of the programme, each of the following needs to be covered:

Theory

Teachers in England are now a particularly well-educated group. To qualify as a teacher one needs a minimum of a three-year Bachelor level degree. Most teachers actually have a fourth, postgraduate, year of professional education. One cannot treat such people as technicians, asking them to perform certain actions in the classroom without providing them with an opportunity to study the theory underlying the actions, to argue about alternative approaches, and to build their own new skills on the basis of ownership rooted in understanding. It follows that sufficient time in the PD programme must be devoted to explicating something of the underlying theory. The theory of CA has been summarised early in this chapter and that is the theory-matter of CA PD, but the same principle applies whatever innovation is being introduced (apart from simple technical skills such as the use of a new piece of software).

Practice

Joyce & Weil (1986) claim that teachers need 30 hours of practice to perfect a new teaching technique. Whether or not this is always literally true, there is no doubt that one cannot change one's skills without a considerable amount of practice and reflection on that practice, preferably with the help of a supportive mentor or colleague. The inservice part of a PD programme can provide opportunities for practising new techniques in a relatively safe environment, and the onservice section provides the same opportunities but in the real classroom context. For CA PD we actually focussed more on the latter type of practice, although the inservice days do provide time to work with new activities and apparatus specific to CA.

Management

Changing a whole science department is as much a management issue as a technical one. Inevitably, faced with an innovation, a department will include both enthusiasts and more reluctant brethren and keeping the whole department on track and learning from one another is not a trivial task. The PD programme must provide some help in the management of this process by providing activities which address various departmental scenarios. Included in this element of the PD content must be some guidance for the CA co-ordinator and head of department in running effective PD within the school since this will become an essential factor in both effective initiation and in long-term maintenance of the innovation.

Technical

Any innovation which involves new curriculum materials carries with it a plethora of technical questions: Where do I get the print materials? How much do they cost? may I duplicate them in my school? How do they fit with the national curriculum or a given text book? What special apparatus is needed and where do I get it / how do I make it? ... and so on. We recognise that such questions are sometimes offered defensively as a shield against the deeper-rooted reflection on practice which is going to be essential, but at the same time they do often reflect genuine concerns which need to be addressed. The challenge here for the PD provider is not to allow the big aims of the programme to become bogged down by too much attention to detail, while at the same time giving participants a fair crack at addressing concerns which are important to them.

Belonging

If you are going to enter the risky business of trying something radical in your classroom, it helps if you feel part of a movement. Even if you are in the unfortunate position of having little overt support in your own school (see chapter 7 for such a case study), it is possible that in the inservice programme, over coffee, over lunch, perhaps in the pub in the evening, one can build a sense of belonging to a group of people who are doing something radical and valuable. This is the sense that can keep you going through difficult times and an effective PD programme needs to build in opportunities for plenty of social, apparently off-task, activity.[1]

THE PROGRAMME

On the basis of these structural principles and ideas of necessary content (but constrained by cost considerations to be discussed below), in July 1991 we sketched out the following programme for CA PD in which the first group of schools were to enrol, to start in September that year.

2 days INSET[2] at the beginning of the school year (September) to introduce the principles and theory of CASE and to allow teachers to experience the first few activities, become acquainted with the materials available, learn about the pre-test to be used in evaluation, and to begin to think about management implications.

2 days INSET in January, after schools have been using the activities for about one term. This provides an opportunity for teachers to share their experiences so far (and often to learn that they are not alone in not having made as much progress as they believe that they 'should' have), to look at the next few activities, and to take the theory and management further.

1 day INSET in May of the first school year. This again provides an opportunity for feedback but on this day the main activity is 'bridging': participants are encouraged to write their own CA-style activities set in the context of topics from the national curriculum. The purpose is for them to use their understanding of the principles of CA – especially the pillars of cognitive conflict, social construction, and metacognition – and to apply them to new teaching situations.

1 day INSET in October of the first term of Year 2 provides a chance to adjust to staff changes with the new year, to consider the implications of starting the activities in Year 8, and often to induct some new teachers into the programme.

1 day INSET in June, toward the end of the whole programme, to go through administration of the post-test and look ahead to maintenance issues, including that of bringing newly appointed staff into the thinking and practice of CA.

5 half-day visits by centre-based CA tutors to each school on the programme, spread over the two years. These are loosely described as 'coaching' visits, but in fact

---

[1] ...but not, perhaps to the extent of a PD day we heard of while writing this, which started at 1000 with coffee and croissants, broke at 1130 for coffee and croissants, offered lunch from 1230 to 1400, then finished the day at 1530 with tea and cakes!

[2] INSET: The Inservice Education of Teachers. A term once in common use in the UK, now generally replaced by the more encompassing term Professional Development. We use INSET here in the restricted sense of a day (or afternoon, or evening, or week) held at a central venue to which teachers come. INSET may include a wide range of activities including lectures, workshops, and practice sessions.

may be used for general introductions, feedback for the whole department, demonstration lessons, team teaching, or for real coaching of various kinds. The value and various roles of these visits will be discussed in more detail below.

(In that first year of CA PD we did propose one additional constraint, driven by logistical considerations, that applicant schools must be within striking distance – say 30 miles - of King's College London. But we had not reckoned with the persistence of one Dexter Hutt, headteacher of Ninestiles School in Birmingham. Dexter was so insistent on the phone that his inner-city school over 100 miles from London, which he had recently taken over and which needed a lot of help, should be part of the programme that we eventually capitulated. That was the start of a long and fruitful relationship which continues to this day.)

After running this programme for two years, we made one important modification, by moving the first two days introduction from the beginning of the school year in which the school was supposed to start using CA methods to the end of the previous year (Figure 3.1). This gives departments an opportunity to prepare themselves for the introduction of CA well before the actual start of the year. This preparation is technical and managerial, but it is also psychological. People need time the muse on new ideas before they try to put them into practice.

This, then, is the basic structure of the professional development programme which we have been running for Cognitive Acceleration through Science Education since 1991. As well as the shift of the first two days back to the previous years, we have continually reviewed the details of the programme and fine-tuned it in the light of experience, feedback, and changing circumstances but it continues to run, starting a new cohort of schools each year. At the time of writing, September 2003, we have recruited 13 schools on to the 13th cohort to go through the programme, and they attended their first two days of INSET in July 2003.

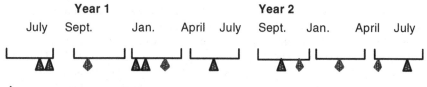

▲ : Centre-based INSET day

♦ : School 'coaching' visit

Figure 3.1: Pattern of activities in a typical CA PD programme

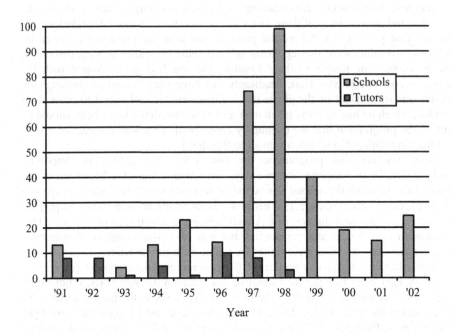

Figure 3.2 Numbers if schools participating in CASE PD at King's College by year

Figure 3.2 shows the number of schools that have participated in CASE PD at King's College London since that first cohort in 1991, but CA is much bigger than this. CAPD programmes have been run at the Universities of Sussex, Keele, West of England, and Chester College. Carolyn Yates now provides all of the training for CASE trainers in Scotland. There are parallel PD programmes for CAME. There are programmes for tutors and for local education authorities and for other groups of schools. Outside the UK there are a wide range of programmes for the professional development of teachers for CA. These all multiply the PD far beyond the King's courses, and each has its own individual features and modifications of our original programme. We will make no attempt to describe each of these in detail, since our main purpose is to present the principles underlying CA PD, to inspect the experiential and academic sources of those principles, and to present empirical evidence for the effectiveness of the programme and its various elements. However, before we embark on this broader process we do need to provide a little more detail of and justification for a number of aspects of the original programme, as well as looking at the significant differences which are entailed in moving CA from secondary to primary schools. This will be the matter of the next chapter.

4. PROFESSIONAL DEVELOPMENT FOR COGNITIVE ACCELERATION: ELABORATION

In this chapter we will look in more detail at some of the features built into, or which emerged in the running of, the original CASE PD programme and then consider the model which had to be developed from it to meet the demands, in some ways very different, of PD in primary schools.

REACHING THE WHOLE DEPARTMENT

We have already noted the futility of trying to change teaching practice by working with just one or two teachers in a school, and our consequent decision to offer the PD programme only to whole science departments. In English secondary schools this typically means something from five to twelve teachers, although a few schools have as few as two science teachers, and a few very large ones may have 20 or more. It is simply not practicable for a school to send all of its science teachers to an INSET day during normal school hours, and it would in any case be difficult for the provider to offer effective workshop activities to all of the teachers from, say, ten schools. Accordingly we advise that each school sends two or three teachers to the INSET days. The importance of having more than one is that they are able to talk to one another, to put the proposals being made to them into the context of their own school, and also that a small team is usually better placed to pass on information and ideas than is an individual.

This practice, however, brings in train another problem, not new to large-scale professional development: the effectiveness of the cascade from those who attend the INSET days to other members of the department who do not. We have found no definitive answer to this problem and do not believe that an all-purpose answer exists, but a combination of the following strategies can provide some alleviation:

1 One member of the department is appointed as 'CA co-ordinator'. This is usually not the head of department but may be a middle-ranking teacher with some ambition. Their tasks include administrative matters such as ensuring that the necessary apparatus is obtained or made, that the relevant worksheets are available at the right time, and that all CA teachers have a clear timetable for use of the activities, but also developmental tasks such as outlining the principles of CA to their colleagues and supporting them as they start to use the activities. The proportion of administrative to developmental tasks will vary with the status accorded to the co-ordinator by senior management and with the inclination and ability of the individual.

2 The INSET days include ideas for those who come about what they may realistically be able to pass on to their colleagues, together with some ideas for PD activities that they themselves can run in their schools. The effectiveness of

this, of course, depends on their being given the necessary time and opportunity when they return to school.

3   The first of the so-called 'coaching' visits by the provider's CA tutors is actually run as a mini-version of the first two day INSET session. This reinforces the key ideas in the minds of those who did attend those days and helps to support their status within the department as well as meeting the more obvious purpose of offering all teachers in the department an opportunity to ask their own questions and raise initial concerns.

On each CA PD INSET day, a school should always send the CA co-ordinator, but the policy for choosing the second (or third) teacher who attends each day varies from school to school (and we do not have evidence to make strong recommendations about this). In some cases, the second teacher is also always the same person, a joint co-ordinator in effect. This has the advantage of providing a second well-prepared person in the event of the first co-ordinator leaving – and being a CA co-ordinator does put one well in line for promotion. On the other hand some schools prefer to send a different second teacher on each occasion so that most members of the department have a chance to experience some of the INSET at first hand.

COST CONSIDERATIONS

In the original planning of the CA PD programme in May and June 1991, we were aware of competing pressures: on the one hand, we were as convinced as we ever had been of the principles that an innovation as deep-rooted as cognitive acceleration demands effective professional development, and that effective PD must (a) be long-term and (b) involve in-school coaching. Inevitably, this must be expensive. On the other hand the UK government of the time had been placing increasingly stringent requirements on universities to offer full financial accountability for their programmes. This meant that we could offer no subsidised training from the university and thus full costs must be passed on to the schools. But there was another policy shift at the time which worked to finesse this potential dilemma, allowing us to design a fully costed high quality programme which schools could actually buy into. That policy was known as LMS – Local Management of Schools. LMS meant that the greater proportion of the funding provided by central government to local government to run the education system must now be passed directly down to the schools themselves. Local education authorities became leaner, offering far less in-house capability for professional development and other services, and the schools became free to purchase services from whoever they wished. CASE as a 'product' based in a high status institution which apparently showed evidence of raising academic achievement was of obvious interest to headteachers with their new-found freedom to obtain the best professional development available for their purposes. This did not mean, of course, that there were no financial constraints. Schools still only had a finite budget for PD each year dependent on the school size and other factors, and most heads were understandably careful to ensure that they obtained good value for money. From the providers' point of view this meant submitting to the discipline of justifying the fees to be charged in terms of both the quality of a programme and its extent.

Inevitably this discipline required us to look carefully at each of the elements in our initially proposed ideal PD programme, and to trim anything that looked like a non-essential luxury. The big problem here is in determining what really is and what is not 'essential'. For one thing, the evaluation of a two year PD programme – especially one that claims to induce long-term effects in students – must take at least three, and more probably five years. Can we wait that long to modify the programme in the light of sound evidence? For another thing, a programme as complex as that of CA PD has so many elements that it becomes practically impossible to attribute successes and failures to particular elements, isolated from others. This of course is a main issue to which this book is attempting to contribute some insight, and in the concluding chapters we will need to discuss the problem in more detail. For now, while discussing the initiation and 'running repairs' to the CA PD programme in its first few years, we had to rely to a large extent on intuition to estimate what really was essential in the programme, and what could be trimmed in the interest of keeping the overall cost reasonable. (Note that intuition may be implicit knowledge, but it is rooted in experience; an expert may not be able to point to chapter and verse to justify a particular decision, but the intuition of an experienced person is certainly worth a great deal more than that of a novice.)

The general message here is that issues of cost are critical in the design and implementation of professional development for teachers. This is true whether or not the provider has to charge the full economic cost or the client has to pay the full cost. Even in systems where costs are subsidised, for example by government or charitable funds, such sources are never bottomless pits and as soon as we start thinking about large scale – say national or state-wide – roll-out then even the most beneficent of funders will start to ask "How much will it cost?" and "Can't you do it cheaper?". The challenge for the provider is to maximise the quality of the programme while both keeping it within the realms of financial possibility for large scale implementation and not compromising essential principles such as the need for in-school follow-up work.

In fact it is not cost *per se* which is critical, but cost-effectiveness. This is still not well enough understood, or at least not well enough acted upon, by schools in the market for professional development. The virtual uselessness of one-off one-day INSET courses, except as introductions to more substantial programmes, is now well-established in the educational literature, but they remain a common form of 'professional development'. This can only be because of a short-sighted attitude amongst those buying into such INSET, an inability to see that money spread around amongst many such days, to give everyone a chance, is actually completely wasted. Far better to pour all of one's allowance into a substantial programme, even if it only reaches a small proportion of one's teachers each year. Headteachers who choose to invest in expensive but effective PD programmes targeted mainly at one or two departments may have to argue that although an apparently disproportionate amount of the PD budget is going to one department this year, other departments will receive similar support in subsequent years.

We have seen how even an expensive two year programme including a substantial in-school coaching component can be financially viable on a large scale and over many years, at least for secondary schools, provided that it offers high quality, theory-based, professional development and continues to demonstrate its effectiveness in measurable gains such as in academic achievement. When we look at the CA programme in primary schools, later in this chapter, we will see that the situation is somewhat less certain.

"COACHING" VISITS

We have made much of the need for coaching visits. As long ago as 1977 Pauline Perry (quoted in Hopkins, 1986 p. 5) wrote:

> "The case has been cogently made that to ensure true implementation of change ... we
> must work with teachers in the place and in the situation where change is to take place."

Empirical evidence for the importance of coaching provided by Joyce & Showers (1995), already referred to, was based on a stringent criterion of 'effective' professional development, accepting only studies which showed significant effects on the learning of students. In other words, studies which only reported changes in teaching practice without concomitant increased student learning were rejected as 'not proven' as effective. On this criterion, it became very clear that the only staff development programmes which were effective were those which included an element of coaching, defined as work with teachers in their own classrooms, as an essential supplement to centre-based course elements. In chapter 11 we will reflect on possible theoretical mechanisms by which coaching operates, in the light of a model of teacher change. At this point, discussing the development of the CA PD programme, it will suffice to consider the face value of coaching in influencing classroom practice, the various forms that it can take, and our own application of the methods within CA PD.

The following paragraphs describe the main activities we engage in on our school visits, together with some reflection on their effectiveness and whether or not they constitute 'coaching'.

Meetings with senior management

The role of senior management in firstly deciding that a school will participate in the PD programme and then providing the necessary support for attendance at inservice days and time for in-school meetings is obviously critical. On our first visit to a school we make a point of meeting with the headteacher or his/her deputy responsible for staff development. This is partly a courtesy, partly an opportunity for them to get some impression of the sort of people we are, but most importantly from our point of view a chance to emphasise again what has already been written in preliminary documentation: that this PD programme is long-term, that it require serious commitment by senior management and by at least key members of the department, and that it will never work like an injection of our expertise into the school, but will work as a process of mutual construction of new approaches to teaching. This first meeting with the headteacher allows us to start to form a judgement about the leadership style of the school. It also covers us against possible future accusations of not having made clear what the commitment of time will be for effective change, and ensures that as far as possible the work we do with the science teachers will be seen to have the active support and backing of the head.

In spite of these precautions, it does still happen on occasion – and it continues to surprise me – that a headteacher can in one month happily release a substantial sum of money for our training programme and then, half a year later, find all sorts of reasons why the teachers cannot be released to participate in the programme. Fullan (2001) p. 35 describes a type of leader as a 'pacesetter', one always charging ahead of his or her staff, looking for the next bright idea. This leadership style is not well-suited to long-term innovations which aim for slow but deep changes. Fullan describes the type of school led by a pacesetter headteacher as 'Christmas Trees Schools', glittering with one innovation after another, pretty from a distance but with little staying power. Luckily our experience of such heads has been very slight.

There is no way in which these initial meetings with heads could be described as 'coaching'. They have a predominantly managerial purpose, but they are an essential, if time-wise small, element in the programme.

Meetings with department

As described above, our first visit to a school typically includes a meeting with the whole science department to run a mini-version of the first two INSET days, presenting the general principles of cognitive acceleration and main aspects of the planned implementation. Most importantly, this gives all members of the department an opportunity to ask questions. Since at a first meeting with the university-based tutor, some teachers are uncertain about what they want to ask (it is all so new, where do you start? might you look foolish?) we split them into pairs and invite each pair to generate at least one question. The minimum duration of such a meeting is 90 minutes. Some schools like to arrange these as twilight sessions, typically from immediately after school at about 3.45. Others make arrangements to free an afternoon for the meeting.

We actually only instituted these meetings after the first couple of years of running the programme, when feedback from teachers who had not attended the central INSET days revealed that many of them felt completely in the dark about cognitive acceleration for months into the supposed initiation of the project in their schools. Clearly we had been over-optimistic about the process whereby those who had attended the INSET days were to explain the programme in some detail to their colleagues. This failure would have been a compound of us not emphasising sufficiently the need or such transfer, the CASE co-ordinators not yet having sufficient confidence to attempt the transfer after just two days of INSET, and lack of opportunity afforded to them by their head of department.

We would not describe these departmental meetings as 'coaching', since they do not occur with teachers in their classrooms with their students.

Demonstrations

Schools frequently ask us to do demonstration lessons with one or more of their classes. We have rather mixed feelings about doing such demonstrations. Physical education teachers and art teachers have a general rule that they do not demonstrate gymnastic feats or do a drawing for their students on the grounds that if they do it perfectly, it will discourage students by setting an impossibly high standard, while if they fail students may well say "well if she can't do it, how can she expect me to?" In the same way, one has to be very careful in teaching a demonstration CASE lesson in front of other teachers. To be sure, it is a lot of fun doing it, taking a class or group of children quite new to you and getting them to start to think in new ways, to talk with each other in a manner to which they are unaccustomed, and to pull out tricks which are part of one's repertoire for getting easy laughs. If this is all there is to it, the demonstration lesson is no more than an ego-trip for the tutor.

But there can be more to it than this and demonstrations can offer models of interaction with students which may be new to the teacher, and may show that a new technique does indeed work in this classroom with these students. We suggest that at least some of the following conditions need to be met:

1   Observing teachers have already made an attempt to teach CA activities, and have some understanding from the inside of what CA lessons are like. This makes it far easier for them to recognise features of the lesson specific to CA and relate their own efforts to the demonstration.

2   Observing teachers are given specific things to look for which are characteristic of CA, such as questioning techniques, ways of involving the whole class, tolerance of uncertainty, or the elicitation of metacognition. To this must be added an opportunity for de-briefing, for the tutor to receive criticism of his or her effort and comment from the teachers on how they would apply those techniques in their own classes.

3   A sufficient number of teachers observe the demonstration. For a demonstration lesson to be seen by only one or two other teachers is simply not cost-effective. (This does not apply in primary schools).

4   (More rarely) when a reluctant teacher has declared that "this will never work with my students".

Under these circumstances, we would describe demonstration lessons as a form of coaching.

Observation and feedback

This is the classic method used by teacher tutors in pre-service teacher education programmes and also (sometimes without the feedback) by inspectors and appraisers. It therefore carries with it quite a lot of baggage associated with being judged, and teachers are sometimes understandably uncomfortable at the prospect of being observed through a whole lesson and then offered comments on their performance. When using observation and feedback as a method of coaching it is therefore necessary to build up confidence that the process comes from a different perspective from that used in the judgmental versions. In practice this cannot be achieved simply by re-assurances beforehand, but can be achieved by the action itself. That is, when the feedback is seen to be supportive, confidential, and focusing on techniques specific to CA it is then generally accepted as useful and welcome. Naturally one applauds good techniques, makes specific suggestions for re-wording or re-timing or other re-emphases where necessary, and encourages maintenance of the process of pedagogical development (a process known amongst some of our primary teachers as 'the shit sandwich': good news, bad news, good news).

Amongst CA tutors a variety of methods is used for recording the action of a lesson. Michael Shayer makes shorthand notes of as much of the dialogue and action as possible, types this up in the evening, and gets it back to the teacher with commentary by the next day. I do something similar but using a laptop in the lesson (which some colleagues suggest is more intimidating to teachers). The advantage of either of these techniques is that the teacher gets an outline transcript of their lesson, with timings, and commentary on specific actions and possible alternatives related to the five pillar model of cognitive acceleration. The disadvantage is that the feedback is not instant and if, as sometimes happens in practice, the transcript is not received until some days after the observation, some impact is lost.

Tony Hamaker also provides an outline of the main moves of the lesson but with a fuller following commentary on the strengths an areas for development, again related to the five pillars of cognitive acceleration. Chris Harrison tends to offer less detail of the lesson but more comprehensive written feedback. Anne Robertson, working in primary classrooms, provides verbal feedback immediately to the teacher, although there are some practical problems with this which we will return to in chapter 9.

What none of us do is to use any sort of observation schedule. Our observations are framed mentally by the principles of CA – concrete preparation, cognitive conflict, social construction, metacognition, and bridging and with reference to the schemata of formal or concrete operational thinking. CA tutors have these so well internalised that they recognise good and bad examples of their realisation and can make judgements about the relative time devoted to each. In the early implementation of CA@KS1, for example, we found that teachers were taking far too long over concrete preparation – a mode in which they were comfortable as it is not too different from regular teaching practice – and so postponing the essential cognitive conflict / social construction phase and running out of time before there was a chance to elicit metacognition. A minor corollary of this approach is that it is useless to try to observe the first half of one lesson and the second half of another. If two CA lessons are happening at the same time, one must choose one or the other. That is the only way that the full picture and pattern of the acts of lesson can be observed.

Figure 4.1 shows a complete set of coaching notes written by Tony Hamaker on one lesson he observed. This has been anonymised, and spaces removed, but otherwise it is unedited, and gives a good impression of the immediacy of the feedback, and of the combination of praise and constructive criticism characteristic of what we believe to be good coaching.

(heading gives information on school, class, teacher, activity, date, and time)
CODE USED: *CP (Concrete Preparation); CC (Cognitive Conflict); CN (Construction); MC (Metacognition); BR (Bridging); T (Teacher); St (Student)*

Time	Observation	Pillar
1325	Class ready to work. 26 in lesson	*Nice start and*
	T explains what the lesson is about. Highlights class rules and student boundaries.	*beginning to*
	T explains that the class will do some simple experiments during the lesson. T bridges (BR) back to previous CASE lesson and asks class what they think a variable is	*CASE lesson by BR back to the first lesson.*
	St *"Lots of different things."*	
	(There appeared to be a little confusion here as to what was required from the class with the use of the word 'variable'. You appeared to realise this and you used a very good strategy in terms of trying to get the class to identify variables in their normal science)	*When posing questions, rather than ask*
	T BR into actual previous NC lesson on acids and asks St to identify any variables from their investigation. A number of St begin to describe what they did in the lesson.	*individuals, try posing*
1330	T asks class *"What was the variable that we changed?"* St *"We used different acids."*	*question and getting*
	T *"Yes, but what were we changing?"* St *"Which one was stronger."* Another St describes that one acid was dilute and one was concentrated.	*pairs/small groups to*
	(Here there is evidence that St will describe events –what they did– but find it difficult to answer or explain the answer to your question. You again deal with this nicely)	*discuss it first then take possible*
	T affirms to class that many St seem unsure as to which things were variables. T asks St to suggest anything that happened. One St observed that the tube got hot when Magnesium added to acid.	*answers*
	(Try paired work/small group work to get St thinking, interacting and explaining things to each other)	
	T continues to challenge class to make links into previous normal science and into CASE lesson 1. T eventually suggests that today's lesson is about variables AND relationships.	
1338	T suggests to the class that the important thing about CASE lessons is NOT writing but what the St can say about the experiment.	*CP here*
	(Your whole approach is really good. You are not threatening; the St have so much respect and are relaxed throughout.)	
	T then explains that she will try to help each group with the term variable and relationship when she visits each group.	
	(Perhaps getting the class to discuss words might have been an even more powerful strategy. Then getting individuals to report back the group ideas/consensus)	

1345	Class begin individual circus of investigations. Three teachers in room supporting individual groups.
1405	Class together as plenary

1405 Class together as plenary

(You gave the class 20 minutes to complete the practical work. When you analyse what they had to do, it was really to collect data. The cognitive demand of this is not too high but 20 minutes was spent collecting data. Try to plan ways of data collection which can cut down on the time that students need. In this way, more time can be given for the actual analysis and evaluation of the data by the student)

More CP and some CN

1407 T now gets class to report back their findings and ideas to the whole class. Asks first group to report back ideas from investigation 4. St identify that there was no relationship between the variables weight and height. T asks St to explain how they found this out- why did they conclude that there was no relationship

This is very nice. It allows for CN and MC – well done

St begin to explain how their interpretations of the data yielded this conclusion.

T asks other St to repeat what the first St said. St repeats.

(This is a very good effective strategy to ensure that St do listen to one another)

T gets another group to report back their findings from the pulley investigation. Relationship identified

Another group report on the volume and height of liquid investigation. St suggests that this was very difficult.

(Name of teacher), I was unable to capture the discussion that you allowed for. Apologies. It was excellent. I have to say that the plenary session that you allowed was brilliant. You allowed groups to present their conclusions and asked them to explain reasons and justify ideas and conclusions using evidence. This was extremely powerful.
The collection of the data was a little too long as this could have been shortened to allow for an even longer discussion/reporting session. In CASE lessons, the collection of data should not take too long as this tends to be the least demanding part of the lesson. Once data has been collected the CASE lesson can really take off as it did in your plenary.
I would strongly urge that members of your team observe you at work with CASE. They can learn much from you.
With your questions, try allowing more small group discussion before taking answers. You will be surprised just how much St really do bring to lessons.
Thank you for allowing me to view you at work doing CASE. Keep it up. You might consider yourself becoming a CASE trainer within Camden.
(Name of Authority consultant), the science KS3 consultant is a brilliant CASE practitioner. Why not use her expertise too.

Figure 4.1: Example of coaching notes provided by Tony Hamaker

Team Teaching

Finally in the range of activities in which we engage on the school visits, we come to team teaching. In many ways this is the most satisfactory way of sharing one's expertise with a teacher. She, the teacher, retains overall control of the class and the time, while you, the tutor, are asked to do specific things – or maybe just jump in, with permission from the teacher – at particular moments to offer an alternative approach or to put into practice a technique very familiar to you but possibly novel for the teacher. Generally speaking it is useful if the relative contributions of the teacher and tutor can be roughly mapped out in advance, but there is also value in the spontaneous offering of a take-over for a few minutes as the need seems to occur. What is important is the discussion afterwards about what you were trying to achieve with a frank appraisal of how successful you were and of alternative approaches that might have been better.

One disadvantage of team teaching is that it is quite difficult for the tutor to make useful notes of the lesson progress while at the same time intervening from time to time. Thus team teaching reduces the opportunity for providing substantial written feedback, although it can be argued that observation of and participation in the tutor's practice provides a deeper level of experience for the teacher than simply reading or listening to verbal feedback.

An overview of CA PD school visit activities

In concluding this section on the school visits we make as part of the CA PD programmes, we should consider the overall aims of the coaching and some practical difficulties that are encountered.

The ultimate purpose of the coaching processes – whether demonstrations, observation and feedback, or team teaching – is to assist teachers to change their practice by explicating and demonstrating the methods of cognitive acceleration and giving them plenty of opportunity to reflect on what they have heard, read, and seen and to consider how they can apply it to their own teaching. Ideally, one would like to offer such experiences to each of the teachers in a secondary school department on three or more occasions during the two year PD programme, but this is clearly impracticable. For even a small department of, say, six teachers, this would involve some 18 coaching sessions and even if one could see three lessons in one visit this would require six visits for coaching alone, let alone the introductory visit. In practice time-tabling constraints mean it is often possible to see only two lessons in one day, and many science departments contain 10 or 12 teachers. Under these circumstances it becomes ambitious to aim even as low as one coaching session per teacher over the two year period. How can the effect be maximised? The answer must lie within the department itself. An extra aim of the coaching process is to demonstrate the methods of coaching and to make them explicit. In other words, we aim during the two year period for all teachers in a department to become more comfortable with having others in their classroom, and to pick up something of the process of giving feedback to others as well as to receiving it. In departments which have implemented CA most successfully there is an ethos of sharing experience through meetings, through mutual observations, and through team teaching. All of this takes time and no one should suggest that it is easy to achieve, especially in an environment which emphasises mechanical 'covering' of a national curriculum and which offers 'strategies' as ready-made solutions which can be delivered to schools through an essentially behaviourist training approach. The point to emphasise here is that five visits by themselves are not enough, that to add visits would make the programme prohibitively expensive, and that there is thus a need to aim for a multiplier effect of the visits as delivering not just coaching, but a shift in ethos in a department so that it can continue the process of peer coaching by itself.

CAPTURING / CRYSTALLISING THE PROGRAMME?

To what extent can the complete two year CA PD programme for secondary schools be made available to others? This is nothing other than the old question of whether a curriculum can be crystallised in the form of print, video, and software resources, to be picked up by any reasonably experienced teacher educator and reproduced as a near exact-copy of the programme as designed and delivered at King's College London. I would like to think that even to write down such a suggestion is to reveal it absurdity, but unfortunately there are still otherwise quite well-educated people (not to mention Departments of Education) who act as if a curriculum can be crystallised and duplicated in this way.

Even more than with a content-rich subject curriculum, the curriculum of a professional development programme can never be fully represented or reproduced as a set of inanimate resources, however sophisticated the software may become. We have made the point already but it is worth repeating in this context of professional development, that education is an essentially social process, that is a process involving human interaction. Every interaction between tutors and teachers is a unique event. Its success - evaluated in terms of the extent to which the teachers start to construct for themselves the methods and beliefs associated with cognitive acceleration - depends on nuanced judgement of the tutor who draws from her/his experience and knowledge of cognitive acceleration actions and words appropriate to the particular occasion. At the same time the tutor is always learning, sensitive to new situations and adding them to her/his set of experiences and so enriching further their ability to respond to further novel situations. The moment that a PD tutor says "I've seen it all, I have a standard answer to any question" is the moment that they should seek alternative employment.

This is not to say that a pack of resources to support the PD programme would not be valuable. Indeed after running the CA PD programme for a couple of years we had collectively developed a large and varied collection of activities which we used in the 7 INSET days of the programme. These were concerned with the underlying theory, with technical aspects of implementation, and with management of the programme in schools. They provided practice in recognising the phases of a CASE lesson and with building new CASE-style lessons, and they offered information and visual metaphors about the pillars of cognitive acceleration and about the schemata of formal operations. In 1994, with financial support from BP, we were able to put these activities together as a loose-leaf print resource, arranged as a series of 10 possible INSET sessions which a school might use over the two years in support of the main CA PD programme. That is, it provided the CASE co-ordinator with copies of the activities which she would have experienced at the King' INSET days, so that she could use them with her colleagues in school. It also acted as a resource pack for all CA tutors. The pack was enhanced by the inclusion of a video tape which contained a general introduction to CASE derived from a commercial television broadcast plus a set of 10 very short clips especially shot by a professional team to illustrate each of the pillars of CASE in real classroom settings.

To return to the burden of this section, there was never any intention that this "King's BP Thinking science INSET Pack" (Adey, 1993) as it was known could ever replace the CA PD programme as run by the experienced CA tutors. We never deluded ourselves that a glossy pack of resources, however expensive to produce, could ever be a substitute for the level of human interaction and mutual growth in understanding that occurs in a real high-quality professional development programme. Apart from anything else, such a delusion implies that at some moment one can crystallise the 'best' programme, as if any good programme must not of necessity be continually developing. Our INSET pack[1] was never intended to be more than a support to CASE co-ordinators who were on the programme and to other CA tutors. In chapter 12 we will consider the dilemma facing a national education department charged with delivering high quality PD on a massive scale, but with insufficient resources.

TUTORS

Where do CA Tutors[2] come from? The last section may give the impression that only a small select band of initiates is qualified to run the CA PD programmes. There is an element of truth in that, but as the programme expands nationally and internationally it is necessary to bring more people on board as CA tutors.

From the start of CASE PD in 1991 we have run, in addition to the programme for schools, a parallel programme for tutors. In the first cohort we had 12 schools and 8 tutors. These tutors were mostly local education authority inspectors / advisory teachers who planned to introduce CASE into their authorities, although one or two were university lecturers who wished to start running their own CASE PD programmes. The tutors attended all of the INSET days for schools and also came in for additional days to reflect on how the INSET days were run and contribute to the model of coaching that we were building. Back in their own institutions, they identified a school where they could themselves start to teach the *Thinking science* activities, and then a set of schools where they could start to offer professional development for cognitive acceleration. The visits which we offered these tutors-in-training were (1) to observe them teaching and provide coaching, (2)

---

[1] This INSET pack is now out of print. At the time of writing we are working to make much of the same material available as a CD ROM. (Adey, Shayer, & Yates, 2003)

[2] In the past we have been a little careless in using the terms 'trainer' and 'tutor' interchangeably. Since the term 'trainer' has a rather instrumental, instructional, ring to it, we will try to stick to the term 'tutor' in this book.

to support them in starting their own PD programmes, running some of the activities and observing them run others, and (3) to accompany them as they visited schools, to support them in the development of their coaching skills. All of these individuals were people with some experience of working with teachers, either in an advisory role or as pre-service teacher educators, and so we were building on existing skills and sharpening them with respect to the particularities of cognitive acceleration (notably, the five pillar model and the schemata of formal operations.)

This programme for tutors continued for some years, during which about 25 individuals completed the course. However, the situation in local education authorities (LEAs) was changing as the Conservative Government policy of Local Management of Schools meant that LEAs had to pass most of the education budget to schools themselves. The schools could use their funds to purchase services wherever they wanted, rather than relying on the LEA to provide them for 'free'. The effect of this, as intended, was that LEAs either lost much of their inspectorate and advisory staff to private companies which were set up to provide services such as professional development, or created their own cost centres which had to justify their existence financially by selling services to schools - outside heir own Authority as well as inside. In this climate LEAs were less inclined to make the investment in training an advisory teacher as a CASE tutor, with the risk that either that individual may subsequently move to another authority, or that they could fail to recoup the cost of training by selling CASE PD to their own schools.

From 1994 the numbers of applicants for the CASE tutor course declined sharply, and we no longer had viable groups for whom we could run dedicated courses. Rather, the two or three individuals each year had to attend the regular schools' programme, and we aimed to meet their special needs by occasional extra half days, sessions apart from the mainstream, and different sort of work in the LEAs. It has to be said that this had mixed effects. In some cases the tutors in training were given adequate support by their authorities and were able to institute and run successful CA PD programmes with their own schools. In others (in a situation parallel to the 'Christmas Tree Schools' mentioned earlier) Chief Education Officers (or Directors of Education, as they came to be known) seemed to think that nominating an individual for the CASE tutor programme and paying the fee was all that had to be done. The individuals were not given time for their own teaching practice with CASE activities (which we always maintained as a non-negotiable element of their training) and seemed to be expected to run CA PD courses for schools (who paid the authority) almost as soon as they had started the training and without further assistance from the central team.

The types of situation which arose can best be illustrated with two vignettes drawn directly from real cases, but of course anonymised. (We would not grace these with the grander title of 'case studies'. We will offer some genuine case studies in chapter 8). Both of these are rural LEAs.

Pennine County

This is a large rural LEA whose schools are relatively small and widely spread. Pennine proposed that Emily Scrimshaw become a CASE tutor and she was accepted on to one of the CASE PD programmes as a tutor in training. Emily was an experienced science teacher who had been head of a department and was now employed by the LEA as an advisory teacher. She attended all of the INSET days (see chapter 3 for the outline programme) and also, in the company of two other tutors-in-training, two extra half days at King's on two of the occasions she was in London for the regular INSET days. In her LEA, soon after starting the two year programme, she identified a school where she could take a Year 7 class once every two weeks to start to teach the *Thinking Science* lessons herself. She maintained this teaching commitment with virtually no interruption for the two years. In April of the first year, she called an afternoon meeting of headteachers and heads of science in the Authority where she outlined the nature of the CASE programme, the benefits as she perceived them, and the commitment of time and money that would be required by a school which wished to start training for CASE. After some further discussion about costing and commitments, eight schools said that they would like to be trained. Emily started with a two-day INSET session in July, paralleling the two days run at the start of the King's programme. An established tutor from King's attended one of these days, made some input, observed Emily's training and offered feedback to her. The schools started to teach CASE in September, with the usual variations in ease of implementation. This was the start of Emily's second year on the King's CA PD course and she continued to teach *Thinking Science* herself in one school, while supporting her cohort of eight schools with visits and more INSET days in the pattern of the King's course, inviting a King's trainer occasionally to make inputs and to observe her on INSET days and when she visits schools. Emily is now an established CA Tutor who participates actively in the annual CA Convention and attends many special 'forum' days run for CA tutors where they can exchange information and consider the place of CA within the constantly shifting sands of government policy initiatives. This, we need hardly add, is a success story.

Gloamshire

This authority has hived off its advisory and inspection services to a private company, EdServices Ltd. They proposed sending two people on the CA Tutors' course: Jeremy Jones and Sarah Carin. Jeremy had been an advisory teacher in the authority and was now fully employed by EdServices. He was obliged to bring in enough work paid for by schools to justify his salary plus the overheads of the company. Sarah was head of science in a small school where she had been a teacher for some 15 years. Jeremy and Sarah were old friends, both having worked together in the same LEA for many years. They attended the first two days of the CA PD course as well as an extra session at which we discussed how they were going to operate together to make CA PD available to schools in their LEA, and beyond if they wished. At this stage Sarah was uncertain that she would be released to visit other schools. Jeremy seemed willing to explore the possibility of finding a school in which to teach the CASE activities himself. At the next INSET session at King's, Jeremy reported that his boss would not let him go to a school for regular teaching as he (the boss) did not see this as a cost-effective use of his time. Sarah had made a good start with the activities and was enjoying them (and teaching them well, as we observed.) but it became clear that she had no intention of leaving her own school and students to offer her experience and growing expertise to other schools. I went to visit the Director of Education to explain the principles of CA PD and put this dilemma to him. He appeared to agree fully with me and assured me that he would ensure, at least, that Sarah could get around to other schools. He had less control over Jeremy's actions since he did not employ him. It did seem to us that the combination of Jeremy's teacher education experience and Sarah's CASE experience just might make for a viable CA PD programme in Gloamshire as they could work well together, but over the following months the situation did not improve. Jeremy started to offer CA PD courses to schools, having never taught the activities himself, and Sarah remained firmly in her own school. We made it clear that we would not recognise the CA courses run by EdServices Ltd. but apart from that were powerless to improve their quality without the recognition by EdServices that their course did not meet minimum quality requirements. Their driving force was commercial rather than educational.

Of the 27 LEAs which opted to send people to become CA Tutors between 1991 and 1998, I would estimate that some 17 fell more into the success pattern of Pennine, 5 into a failure pattern of one sort or another, and the remainder have been either partially successful or I have no data on them. In the next section we will provide some quantitative evidence for our claim for 'success'.

AUTHORITY-BASED SCHEMES

With the changes in arrangements of government funding for professional development (amongst other things) and the ensuing change in pattern of those wishing to become CA tutors described above, we decided to take a new approach to reaching beyond the limited number of schools to which we and the other university and independent CA trainers could offer CA PD each year.

From 1999 we instituted a system of LEA-based training, which combined the process of training CA PD tutors within an LEA with some direct training of the authority's schools. It works like this: we run the regular seven days of INSET over two years for a group of schools (typically between 5 and 10) in the authority itself. The LEA may have a professional development centre, or may use one of the schools. The authority identifies one (sometimes more) individuals who are to become CASE tutors for the authority. As well as attending all of the INSET days and teaching the activities, they receive extra training in the process of delivering

INSET and coaching. It is these LEA tutors' responsibility to provide the five coaching visits to each school in the scheme. On one of these visits to each school, the LEA tutor is accompanied by an experienced tutor who initially demonstrates the coaching method and then gradually withdraws to a role of observing the LEA tutor coach and providing feedback. Table 4.1 lists the LEAs which have opted to implement CASE PD in tutor- or authority- based schemes.

This scheme works out slightly cheaper than having the schools come to King's, but the main advantage is that the authority builds up its own capability to maintain CA PD indefinitely with only occasional top-up support required from the university base after the initial two years. This at least is the principle. It has worked strikingly well in many authorities, and less well in others. What can we say about the characteristics of implementation in these LEAs, as contrasted with others which have not yet demonstrated such levels of success? Clearly the role of the LEA CA tutor is critical, and the first factor which leaps out from successful authorities is that the CA tutors have been given time both to develop their own understanding and skill and to visit schools to share and to learn with others. Amongst less successful authority-based implementations it is common for a teacher (or two) to be identified as the potential CA tutor but without attention paid to how they are to be released from enough of their current duties to actually carry out the tutor role effectively. These are often excellent teachers with some maturity and, as far as one can tell, full of potential as PD tutors but the release of a teacher from their current duties, even for as little as one day a week, is bedevilled by practical problems. The financial manipulation - how does the school receive compensation to enable them to buy supply or otherwise cover the absent teacher? - is just one. As an experienced teacher the individual is likely to have responsibility for examination classes which she or he is reluctant to leave, as well as other departmental and school responsibilities which now have to be squeezed into four days. Their timetable has to be arranged to free up one day but if it is the same day each week, that may constrain the classes they can observe in the schools they are visiting. And what is the effect on the individual's promotion prospects within the school? The tutor role may enhance their market value if they seek a new job but it will not necessarily endear them to the headteacher when he/she is seeking a candidate for new responsibilities within the school.

Successful authorities have all, from the start or after discovering the problems, either released a teacher full time to become a CA Tutor, moved an advisory teacher into the role, or advertised and appointed someone new. In other words, they have been prepared to make the necessary commitment of time (and that means money) to ensure that the CA PD programme is implemented as thoroughly as it needs to be if it is to have any effects at all.

Table 4.1 Local Education Authorities which have implemented CASE PD

Authority	Cohort	Location	Schools	Tutors
Barking and Dagenham	1998-'00	King's		1
Barnsley	1999-'01	Authority	12	1
Belfast Education and Library Board (ELB)	1995-'97	King's		
Birmingham	1997-'99	Authority	23	2
Blaenau Gwent	1997-'99	King's		1
Bournemouth	1995-'97	King's		1
Camden	2001-03	Authority	5	3
Cardiff	1997- 99	Authority	18	2
Carmarthenshire and Ceredigion	1997-'99	Authority	12	2
Dudley	1994-'96	King's		1
Durham	1992-'93	King's		1
East Riding	2000-'02	Authority	12?	
Enfield	1997-'99	King's	4	1
Gwynedd	1997-'99	King's		1
Hammersmith and Fulham	1997-'99	King's	8	
Hillingdon	2002-04	Authority	6	1?
Islington	2000-'02	Authority	8	
Kent	1991-'93	King's		1
Kingston upon Thames	1999-'01	King's	6	
Lincolnshire	1998-'00	King's		1
Leicester	2001-03	In EAZ	7	1?
Lewisham	2001-03	Authority	7	1
Manchester	1991-'93	King's		1
Newport	1998-'00	King's		1
Norfolk	2000-'02	Authority	12	1
Northamptonshire	1991-'93	King's		1
Northeast ELB	1995-'97	King's		1
Nottinghamshire	1992-'94	King's		1
Richmond	1997-'99	Authority	8	
Solihull	1997-'99	Authority	13	1
Somerset	1998-'00	Authority	8	1
Sunderland	1991-'93	King's		1
Torfaen	1999-'01	King's		1
Western ELB	1998-'00	King's		1
Vale of Glamorgan	1999-'01	King's		1
Westminster	1995-'97	King's	9	

PROFESSIONAL DEVELOPMENT FOR PRIMARY SCHOOLS

In most of the material presented in chapter 3 and chapter 4 thus far, we have been focusing on the long-running professional development programme for cognitive acceleration through science education (CASE PD) and the elaborations of this programme which have occurred over ten years and more. As will be seen from Table 3.1, Cognitive Acceleration is now much more than CASE. In this section we will look at some of the issues which arise when we look, not at PD for secondary teachers working in a department, but at PD for primary teachers who are class teachers.

CA@KS1 is a project which was initiated in 1999 in the London Borough of Hammersmith and Fulham, who had obtained a Single Regeneration Budget from the government to help them regenerate a run-down inner city area known as The White City. They chose to use some of this money to explore the potential of accelerating the cognitive development of the youngest children in the compulsory school system, the 5 and 6 year olds in Year 1. We have described the research phase of this project and the effects on the cognitive development of the children in some detail elsewhere (Shayer & Adey, 2002) and the materials for schools is published as *Let's Think!* (Adey, Robertson, & Venville, 2001). Here we will focus on the PD associated with this material.

Some of the principles remain the same as for the CASE PD programme described already:
- the programme must include centre-based INSET days and work in the schools themselves;
- the INSET days should include some psychological theory, some technical input, managerial support, and plenty of opportunity for feedback and the building of a community of teacher-learners;
- coaching visits will include demonstrations, observation and feedback, and team teaching;
- tutors must have experience of Year 1 classrooms and of teaching the *Let's Think!* activities.

There are some surface differences: we are dealing here with the schemata of concrete operations, not formal, and the activities themselves are, of course, quite different. But there are also some more fundamental differences between the Year 1 and Year 7/8 programmes.
- *Let's Think!* activities are delivered to just six children, while the rest of the class get on with other work. A different group of six is taken each day of the week. Thus the managerial issues are quite different.
- *Let's Think!* is a one year programme, not two years like CASE and CAME. It follows that the PD programme is also only one year. Given (Joyce & Weil, 1986)'s strictures (and our own experience) about the length of time it takes to make real changes in a teacher's practice, there is a question about how much can actually be achieved in one year. In CASE PD it is often during the second year that we begin to notice significant change in many teachers' practice. What is more, it is not realistic to try to fit seven INSET days devoted to one

programme (CA) into one school year - primary teachers simply cannot be released from their schools for that long without upsetting children and parents.

- On the other hand (i) primary teachers are already more education-oriented than are their secondary counterparts. Their own pre-service education has usually had far more learning psychology in it than has that of secondary teachers, whose concerns tend to focus more on pedagogical content knowledge (Shulman, 1987); and (ii) PD for primary teachers works directly with all of the teachers involved - just one or two class teachers per school - and not through even the mini-cascade involved when those secondary teachers who attend the INSET days have to share their understandings with colleagues at school.

- Finally, and by no means least importantly, there is the issue of cost. Secondary schools have budgets for PD proportional to their total numbers of teachers and students and as mentioned already they can juggle these about to provide substantial sums for one department in one year. Primary schools have far fewer resources and are far more sensitive to the charge of inequity if all they have in one year is concentrated on the class teacher(s) of just one year group.

During the development phase of the project we had the luxury of excellent funding and a staff of researchers who spent a great deal of time in the project classrooms. Although technically these were often research observation and data collecting visits, informally they often contributed to the PD effort and enabled the teachers to become very familiar with us and with what we were trying to achieve both theoretically and practically. A strong sense of a community of teacher- and researcher-learners was established. Now we are in the phase of dissemination, and have to face a new sort of reality, that of an enormous demand from schools and from LEAs for the materials and methods. At the time of writing we know of nearly 1000 schools which are adopting *Let's Think!* with some appropriate PD, plus many more which have simply bought the materials. Meeting the demand for PD while ensuring continuing high quality is the sort of challenge that most PD providers would be delighted to have to face! We have proposed that the minimum architecture for a PD programme for *Let's Think!* should consist of three individual INSET days, front loaded with two in the first term of the year and one in the second term, two twilight sessions (one in term 2, one in term 3), and a minimum of three school visits, all to take place over one year while the teachers are starting to use the activities.

A novel plan to build PD for *Let's Think!* systemically into one London borough – the borough that originated the scheme - will form the subject of chapter 9

OTHER CA PD PROGRAMMES

Even now we have far from exhausted the varieties of CA PD which are currently on offer in the UK and internationally. The CAME PD programme managed by Mundher Adhami, whilst having many similarities to that of CASE, does have some distinctive differences. The CASE PD programme in Scotland run by Carolyn Yates is nearer to a 'tutor only' programme and makes good use of video evidence of practice. In the Northeast of England, Reed Gamble's approach to CASE PD relies even more heavily on video evidence from classrooms. In the new (at the time of writing) CASE@KS2 project (*Let's Think Through Science!*), Natasha Serret and others are currently developing a PD programme in two London boroughs which will be made more widely available from September 2003. All of these developments are important and their variations will feed usefully into our general understanding of the PD process, but there is no need to elaborate further on them here. Rather, we propose now to turn to the evidence we have of the effects of the CA PD programmes which have been described in the last two chapters.

PART 2:   EMPIRICAL EVIDENCE

5. MEASURABLE EFFECTS OF COGNITIVE ACCELERATION

In this chapter we will present evidence for the effects of cognitive acceleration on students' cognitive growth and academic achievement. Since, as we have made clear, "cognitive acceleration" comprises a somewhat complex approach to teaching which is introduced over a two year PD programme, such evidence will to some extent be a reflection of the success of the PD. By itself it cannot be conclusive about the effectiveness of the PD since it is at least conceivable that the effects could be obtained without the PD programme or with a less extensive programme than the one we offer. It is however a minimum requirement that we show that CA does have such effects for if it did not, then we could be certain that the programme, including the PD, was *in*effective.

There are two phases of effects of the CASE (*Thinking Science*) programme which have already been extensively reported and which we will just summarise here. But we will add to it some evidence of particular relevance to the evaluation of professional development, some from an apparently anomalous effect and some from individual class (i.e. teacher) effects. We will then summarise the parallel data which we have from the primary level CA@KS1 (*Let's Think!*) experiment, both the larger scale experimental – control differences and, more germane to the present enterprise, the class (teacher) level data.

EFFECTS OF CASE, 1: THE ORIGINAL EXPERIMENT

The original CASE experiment was conducted over three years, 1984-87, using a quasi-experimental method with 10 CASE classes and 10 matched control classes. Of these 10 pairs of classes, four started CASE in Year 7 (designated the '11+' group) and six in Year 8 (designated '12+' group). Tests administered to all students in CASE and control classes are shown in table 5.1.

All test data was analysed in terms of residualised gain scores (Cronbach & Furby, 1970) which show the change scores of the experimental group from the pre-test over and above the change that has been achieved by the control group. Overall the CASE groups made significant cognitive gains at the post-test, although this was largely concentrated in the 12+ boys. There were no differences in science achievement at immediate post test, but in all subsequent tests those students who had experienced CASE in Years 7 and 8 or in Years 8 and 9 outperformed controls. The effect was mostly concentrated in girls who started at 11+ and boys who started at 12+, with effect sizes from 0.32 to .96 standard deviations. Full details of the testing procedures and analyses of the results are provided in Adey & Shayer (1993, 1994); Shayer & Adey (1992); and Shayer & Adey (1993). (Adey and Shayer 1993 is reproduced in Desforges & Fox, 2002). The main feature to note here is that the two-year CASE intervention introduced through a PD programme and set of printed activities with some associated apparatus appeared to produce effects in students which lasted up to three years after the end of the intervention and which transferred from the science context in which they were placed to both mathematics and English results.

Table 5.1: Test schedule in the original CASE experiment

Date	Occasion	Test	Type
9/85	pre-test	PRT[1] II Volume and Heaviness PRT III Pendulum	Level of cognitive development
(CASE intervention: 30 activities over two years)			
7/87	post-test	PRT III Pendulum PRT VIII Probability science test	Level of cognitive development science achievement
7/88	delayed post test	PRT VIII Probability science test	Level of cognitive development science Achievement
7/89	long-term test for 12+ group	GCSE science GCSE Mathematics GCSE English	Nationally set, externally scored school-leaving examinations.
7/90	long-term test for 11+ group	GCSE science GCSE Mathematics GCSE English	Nationally set, externally scored school-leaving examinations.

In this original experiment, there appeared to be little to distinguish the gains made by pupils across the 10 CASE classes. That is, there was no marked teacher effect and at that time we concluded that 'the CASE effect' was relatively independent of the teacher (see Desforges & Fox, 2002 table 8.3 page 200). On the other hand, during that research phase we were working closely with each of these teachers who were the only teachers in their schools using CASE, and so there was no attempt to achieve or to measure the ability of these teachers to transfer their expertise to others within their departments. This was to become an important feature of the PD programme that was subsequently developed, described in chapters 3 and 4.

---

[1] PRT stands for Piagetian Reasoning Task. These are also known as Science Reasoning Tasks. They were originally developed by Shayer and colleagues in the 1970s to conduct a large scale survey of the levels of cognitive development of the school population of England and Wales (Shayer, Küchemann, & Wylam, 1976; Shayer & Wylam, 1978).

EFFECTS OF CASE, 2: SUBSEQUENT ANALYSES

Having reported the effects described in the last section, we now found ourselves in a difficulty over subsequent evaluations. It would no longer be ethically acceptable to use control groups, for to do so would be specifically to withhold from some students an opportunity for cognitive acceleration and so consciously to put them at a disadvantage. Accordingly our subsequent analyses relied on 'natural' experiments, comparing schools which had opted to follow the CA PD programme with either national norms for cognitive development which Michael Shayer had established in the 1970s (Shayer et al., 1976; Shayer & Wylam, 1978) or with schools for which we had data but which were not (yet) participating in CASE. The latter method, in particular, overcame the two objections that (1) norms might have changed between the 1970s and the 1990s and (2) schools which opted for CA PD may systematically be more enthusiastic and stimulating than those which did not. Our 'control' schools were actually also schools which had joined CA PD but whose cohorts of students using CASE had not yet reached the age at which they contributed to post-test measures.

Figure 5.1 illustrates the method used and a typical effect. Each point is one school. The x-axis shows the mean level of cognitive development of students entering the school at the start of Year 7, expressed as a percentile based on national norms. The y-axis shows the mean grade achieved by students in that school when they take their GCSE examination (in this example, in English) five years later, at the end of their Year 11. Unsurprisingly, there is a strong relationship between the average ability of each school's intake and their success at GCSE but what emerges clearly is that CASE delivers a strong 'value-added' effect. Whatever the intake level of the school, if their students participate in cognitive acceleration in Years 7 and 8 the mean GCSE grades in Year 11 are something like a grade higher than would have been expected.

Collecting and analysing this data is not a trivial activity, and we are not able to do it every year. We have analysed and reported data from two cohorts of schools which have followed the CA PD programme described in chapter 3 and 4 – those who started the programme in 1991 and in 1994 respectively (Shayer, 1996, 1999a, 1999b). For each of these we reported effects on the Key Stage 3[1] national curriculum tests (taken at the end of Year 9) and on GCSE in science, mathematics and English. In every one of these 12 analyses (2 cohorts x 2 post-testing occasions x 3 subject areas) there is clear evidence of a value-added effect of cognitive acceleration in every one of the schools for which we have data.

---

[1] The 11 years of Compulsory education in England, from 5 to 16 years of age, is divided into four 'Key Stages'. Key Stage 3 is Year 7 to Year 9, typically the first three years of secondary school.

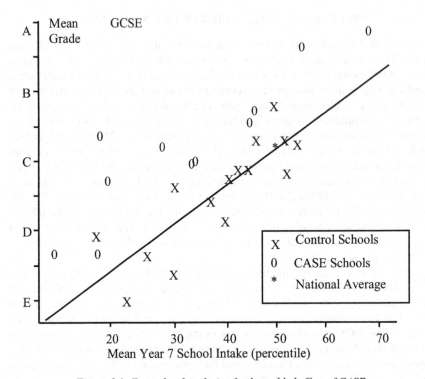

Figure 5.1: Example of analysis of value-added effect of CASE.

So far, this summary of effects of CASE serve only to show that CASE "works", but in a book on professional development we need to be able to attribute this effect, at least in a significant part, to the PD. This will be the enterprise of chapters 6 – 9, but here we can get some insight into the PD effect by looking at (1) a slightly strange phenomenon we discovered in 2001, and (2) the differential effect, within schools, on different teachers who have participated the PD programme.

A STRANGE EFFECT

The full analysis of the 1999 GCSE results (one sample of which is shown above) was completed by Michael Shayer early in 2000. As is our wont, we launched his report (Shayer, 1999b) from the Centre for the Advancement of Thinking with a press release and a bit of a fanfare. With one important exception, the educational press either ignored it (something sexier was probably on the agenda) or published the results without questioning them. The exception was John Clare, education correspondent of *The Daily Telegraph*, who had followed our work for many years and had been generally supportive of our approach to 'raising educational standards'. When he read this report he contacted us with a number of pertinent questions. The detail need not concern us here, but in the process of providing a characteristically comprehensive reply, Michael uncovered a curious anomaly in our data.

Suppose that the mechanism by which cognitive acceleration worked was as simple as this: CASE intervention starting in Year 7 classes in 1994 stimulates the development of students' general intelligence and they are then able to apply this improved intelligence to all of their subsequent learning. This model explains satisfactorily both the long-term and far-transfer effects summarised above. If this was all there was to it, then when we looked at the mean GCSE grades of a particular school (or set of schools which started CASE at the same time) over a number of years, we might expect a pattern as shown in Figure 5.2.

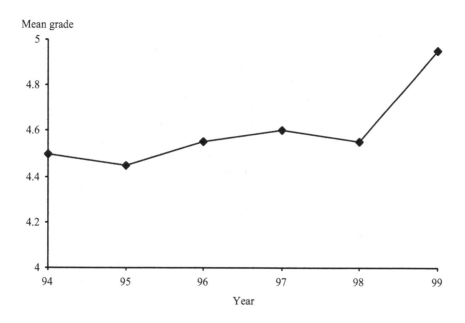

Figure 5.2: Hypothesised mean GCSE grades of a school introducing CASE in 1994

In other words, the GCSE grades would be expected to rise sharply in the year in which the cohort of students who have experienced CASE came through to their GCSE. In fact, the pattern which Shayer uncovered as he worked to answer John Clare's questions is that shown in Figure 5.3.

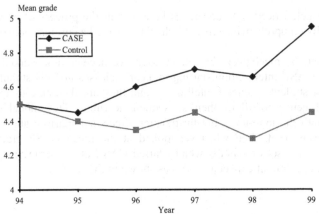

Figure 5.3 Actual mean GCSE grades of CASE and control schools, 1994 – 1999 (science)

For the control group of schools, mean GCSE grades fluctuate slightly but without significant differences from year to year. But for the CASE schools the GCSE grades do not rise sharply in 1999 as anticipated, but rather rise gradually from two years after the CASE programme is introduced to the schools. The students sitting GCSE in 1997 and 1998 have not themselves experienced CASE, so what is the explanation for their improved grades? The most plausible hypothesis we can offer is that the effect is as much to do with change in the teachers' general methods of approaching the process of teaching and learning as it is to do with the specifics of the CASE activities. In other words, the CA PD programme addresses, as it is intended to, teachers' implicit beliefs about the nature of learning as well as their particular skills in questioning (and in pausing) to induce cognitive conflict and metacognition and this has an impact on all of their teaching. Within a year or year and a half of starting to participate in the PD programme they are stimulating all of their students, not only those in the years which are using the CASE activities.

A similar conclusion may be drawn from an observation made by a number of commentators on our results, that it seems unlikely that such large long-term far transfer effects could be achieved simply from the substitution over two years of 30 regular science lessons by 30 'special' CA lessons. The effect must be more diffuse and pervasive, and this is just what one would expect from changes in teachers which influenced all of their teaching.

A further aspect worth noting is that a very similar effect of gradually rising GCSE grades is observed in GCSE English as well as in science. To explain this we need also to suppose that the stimulation provided by the CASE-trained science teachers has a very general effect on the information processing capability of their Year 10 and Year 9 students, which the students then apply – probably unconsciously - across the curriculum. The relevance of this effect and speculation to a model of educable general intelligence will be clear, but a book on professional development is not the place to pursue it. The interested reader is referred to Adey (1997, 2004); Adey & Shayer (1994).

Class level effects from CASE

So far we have been considering only whole-school effects, but if the PD programme we offer is to fulfil our hopes for it, it must be shown have an effect on a majority, if not on every, teacher in the departments with whom we work.

Figure 5.4 shows the cognitive gain made by each class in the eight schools which participated in one of the cohorts of Cognitive Acceleration through science Education Professional Development. This cohort is selected for display simply because the data is readily to hand and unusually complete. Each bar represents one class. The classes are almost arbitrarily labelled, except that '7.1' means the first class in school number 7, and so on, so one can see the number of classes in each school. The length and direction of a bar shows the mean residualised gain in levels of cognitive development made by the students in that class – 'residualised' in the sense that it is the gain made over above that made by a control group, so that a negative value means that that class actually did not develop as much as did the control group. These are expressed as effect sizes, that is, raw residualised gains divided by the standard deviation of the control group gain. For effect sizes, 0.35 is normally considered satisfactory, and 0.5 good. What is striking from this figure is that out of 63 classes whose teachers participated, either directly attending the INSET days or indirectly through in-school-based INSET and coaching visits, all but five achieved gains greater than the control mean. To be sure, not every class that has positive gains will have reached statistical significance, but the overall picture is one of very wide-spread gains including all but one (school 7) of the eight schools involved. We believe that we are justified in claiming this as evidence that the programme is indeed reaching into the schools to teachers beyond those who attend the INSET days.

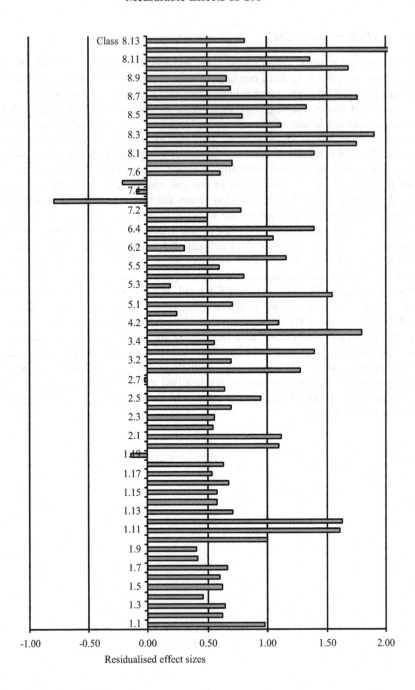

Figure 5.4: Residualised gains of each class in one cohort of CA PD schools

EFFECTS FROM CA@KS1

CA@KS1 is an acronym used for CASE at Key Stage 1[1], the project introducing cognitive acceleration to Year 1 children in the White City area of Hammersmith and Fulham described in chapter 4. The materials of the project have been published as *Let's Think!* (Adey et al., 2001) and this is how it is now generally referred to.

This project differed from the original CASE in some important ways in that it was:

• aimed at 5 to 6 year olds (not 11 – 14 year olds)
• based on the schemata of concrete operations (rather than formal operations)
• designed as a one year intervention (not two)
• intended for use by class teachers typical of primary schools (not subject teachers typical of secondary schools).

As a completely new project, the effect of which was quite unknown to us, we had no compunction in running a new quasi-experiment. 14 Year 1 classes in 10 schools were provided with the material and their teachers attended the one year PD programme outlined in chapter 4. A further 8 classes in 5 nearby schools were identified as controls. The testing programme was as shown in table 5.2.

Table 5.2: Testing programme in the CA@KS1 experiment

Date	Occasion	Test	Type
9/99	pre-test	Drawing (spatial ability) Conservation	Piagetian measures
		(Let's Think! intervention: 30 activities over one year)	
7/00	post test	Drawing (spatial ability) Conservation	Piagetian measures
7/01	delayed post tests	Raven's coloured matrices NC KS 1 tests	non-verbal intelligence Mathematics and English achievement

Over the one year period of the intervention the CA classes showed significantly greater cognitive growth than the control classes, with effect sizes from 0.35 to 0.59σ. These quantitative results have been reported in detail in (Adey, Robertson, & Venville, 2002). As with the overall effects shown by CASE, these gains do not themselves provide direct evidence of the effectiveness of the PD programme although they are a necessary precursor to establishing the effect of the PD. However, two qualitative studies of what actually happens in Year 1 CA classes throw considerable light on the type of thinking and dialogue which become typical of CA classes, whether or not they are engaged in cognitive acceleration activity.

Grady Venville (Venville, 2002; Venville, Adey, Larkin, & Robertson, 2003) showed that there was significantly greater use by children in CA classes compared with controls of a number of important strategies related to the development of

---

[1] Key Stage 1 is Years 1 and 2 (equivalent to US K and grade 1)

thinking, such as explaining, highlighting a discrepancy, adopting a new idea, and working collaboratively. Anne (Robertson, 2002) showed that when children in CA and non-CA classes were asked what helped them to learn in numeracy lessons, CA children typically answered "listen", "talk", "discuss", "explain", while those in non-CA classes emphasised "be quiet", "copy", "don't be naughty". Both of these studies show the impact of the CA programme on the development of social construction and metacognition, precisely the types of student skills on which the PD programme focussed.

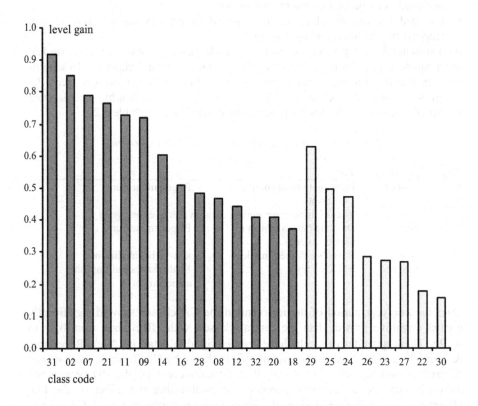

Figure 5.5: Residualised Cognitive Gains by Class; CA classes on Left, Control on Right

When we look at the mean residualised gains made by the classes of particular teachers (Figure 5.5), we can see directly that although generally the CA classes showed greater gains than did the control classes, the effectiveness of teachers varied considerably. This is much as one would expect as individuals take more or less from the PD depending on their experience, on their starting implicit beliefs in the nature of teaching and learning, and on their ability to comprehend the nature of teaching for cognitive acceleration and to change their practice accordingly. The class data also reflects the unsurprising fact that a few teachers in control classes already appear to offer good cognitive stimulation, although none of their cognitive gains approach the best of the CA teachers'.

I (PA) did the quantitative analysis of this pre- to post- test data by class but before sharing it with my two research colleagues on the project, Anne Robertson and Grady Venville, I asked them to rank the 14 CA teachers according to their opinions, based on many hours of co-operative working and observation in the classes over the year and on intense participation in the professional development process, in terms of their opinions of the extent to which each teacher had internalised the methods of cognitive acceleration. Initially I was encouraged by their responses since the rank correlation of their judgements was 0.94. However, when I then compared their rankings to the cognitive gains made by the teachers' pupils, the best correlation between the researchers' judgements and the measurable effect was only 0.35.

What can we conclude from this? Firstly, that the three of us deeply involved in developing the materials and the PD programme for CA@KS1 had, within 18 months, formed a remarkably consistent construct of what counted as effective teaching for cognitive acceleration. Secondly, that the relationship between this construct and effects in students is not straightforward. It might be argued that features we highlighted in our construct, although consistent amongst us, are not actually those most important for cognitive stimulation. But those features, such as long pauses for children to construct their thinking, the careful establishment of excellent social-cognitive relationships within the pupil groups, and the encouragement of metacognitive reflection on the type of thinking the children had been using, all arise directly from the theoretical model of cognitive acceleration on which all of the work is based and which has provided clear evidence of cognitive effects in CA groups as compared with controls. Bear in mind that I did not ask Anne and Grady to use specific check-lists or observation schedules which highlighted these features, so there may well have been something of a halo effect on the judgements of individual teachers, perhaps with more experienced and apparently confident teachers being over-rated and younger apparently more nervous teachers under-rated. Certainly, however, a major contributor to this low correlation would simply have been the complexity of the process of relating particular teaching procedures to particular cognitive outcomes. We will be discussing the notion of process-product research in more detail in chapter 10 but here we need only re-iterate the observation frequently made in connection with assessing the effects of teaching on learning, that there are so many mediating variables involved that it is quite unrealistic ever to expect clear, lock-step, relationships to emerge.

DELAYED EFFECTS?

From the testing programme outlined in table 5.2, and in parallel with our evidence-collecting from CASE, it should be clear that we were looking for long-term effects of the Year 1 cognitive acceleration programme on the children's development and achievement. Up to the time of writing, data from delayed tests have failed to bolster this hope. One year after the end of the one-year CA intervention in the Year 1 classes, there were no longer any significant differences between experimental and control children on either general intelligence tests or on mean achievement scores on national curriculum tests in English and mathematics which all children take at the end of Year 2. We looked separately at boys and girls and those who started from relatively high and relatively low levels at pre-test, but the only positive effect we discerned was that a significantly higher proportion of children who had experienced *Let's Think!* achieved the highest level (3) on the mathematics Key

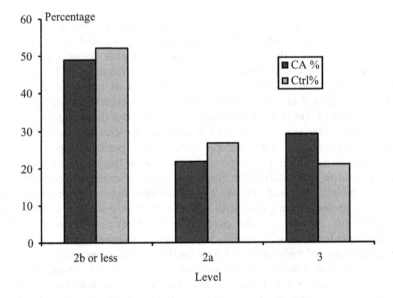

Stage 1 test than did control children (see figure 5.6).

Figure 5.6: Percentage CA and control children achieving levels at Key Stage 1 Mathematics test.

Another interesting effect was that there was a correlation of 0.52 between the class mean gains on the conservation immediate post test and on the delayed Ravens matrices residualised gain scores, across all 22 CA and control classes. For the CA classes alone, that correlation reaches 0.64. Corresponding figures for the mathematics achievement test are 0.41 and 0.54, and for language 0.26 and 0.35. In other words, there is a trend for classes who have made larger cognitive gains over the period of the intervention also to retain longer-term effects, at least in measures of non-verbal intelligence and mathematics. Although it appears that this trend is not strong enough to maintain an overall significant effect of cognitive acceleration, it is suggestive. Our data actually shows that two of the control classes, from the same school, made extremely large gains in the year after the intervention, that is the year leading up to the national curriculum Key Stage 1 tests. That we know this school to be very examination-oriented, one which puts great store on public achievement measures, offers us no excuse since it could be argued that if examination success can be achieved by traditional cramming methods, why bother with the subtlety and expense of cognitive acceleration?

It is worth going back here to ask why we opted for a one year intervention in Year 1, when we had insisted on two years as a minimum for the CA work with 11 – 14 year olds. At the time the reasoning was that one year is nearly 20% of a five-year-old's life, that their brains are probably more plastic at that younger age, and with a special activity every week we should have a good chance of having a permanent effect on their cognitive development. A further consideration was that we were working directly with the teachers, not through the one-step supported cascade from CASE co-ordinator to the rest of the department characteristic of the Year 7/8 work.

With hindsight, and in the absence of good evidence for a long term effect, we should re-visit this reasoning. On brain plasticity it probably is true that one can have a quicker effect, but there is another side to the coin: without continuing special stimulation the developmental process may revert rather quickly back to the norm. We suspect, however, that a critical factor is one more closely associated with the theme of this book, that is, it is to do with the professional development of the teachers. There are two problems here: Firstly, we worked with the Year 1 teachers for only one year. Joyce & Weil (1986) have claimed that new teaching skills require at least 30 hours of practice. Whether or not one fully buys into this claim, our own experience with CASE shows that it often not until the second year of the PD that teachers start to feel comfortable with their new pedagogical 'clothes'. Our Year 1 teachers never had the chance, during this experiment, to revisit their new practice for a second year. Secondly, the Year 2 teachers were generally unaware of the methods of cognitive acceleration, so the children experienced one year of stimulation followed by one year of normal – probably good quality – teaching for recall and concept development. Even where a trained CA teacher moved up with her class to Year 2, there were no special CA activities, no special time allocated to cognitive stimulation, and often considerable pressure exerted by the looming Key Stage 1 national tests. (Anyone outside England must be stunned at the thought that children as young as 6 are already being subjected to examination pressure. Does it make the English a particularly intelligent and thoughtful people? A brief look at some of our tabloid newspapers will provide a quick answer to that).

We have said nothing in this chapter about the influence of headteachers. This is not because we consider it unimportant, or because we did not notice any such influence. On the contrary, in primary schools it is often more immediately obvious than in secondary schools. We will be dealing with this issue in chapter 9 where a full account of the systemic professional development programme introduced in 2000 – 01 will be described. Here, we need only note that the cognitive gains made by the children in that programme were comparable to those reported here for the

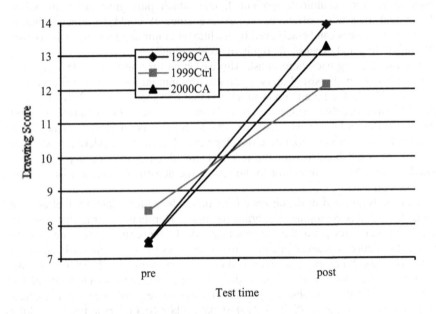

1999 – 2000 programme, as is shown in Figure 5.7.

Figure 5.7: Cognitive Gains of '99/'00 and '00/'01 CA@KS1 Cohorts Compared with Controls

This work continues, and our current thinking is that CA@KS1 must become a *two-year* programme. With this in mind, Anne Robertson in Hammersmith has and trialled some CA activities for the Reception class (4 to 5 year olds) and, with Michael Shayer and Mundher Adhami, also developing CA activities for Year 2.

6. TESTING AN IMPLEMENTATION MODEL

I would like to thank Professor Margaret Rutherford late of the University of Witwatersrand, Ulrike Burrmann (graduate student of the University of Potsdam), and Sarah McGlinn (undergraduate psychology student of Middlesex University), for many hours of help in interviewing, cross-validating, and (Sarah) data-entry related to the studies reported in this chapter. I am also grateful for the extensive work put in by my colleagues at King's, Justin Dillon and Shirley Simon, on analysing the interview data from headteachers and heads of science.

We have emphasised that the critical outcome of the CA programme should be a measure of student gains in levels of cognitive development and of academic achievement and in Chapter 5 we presented a range of (mainly) quantitative evidence for such gains, as well as some data from classroom observations and from class-level effects which are suggestive of the role that the CA PD programme plays in developing teachers' practice. Now we will offer a more complex study in which we explore a number of variables which seem likely to influence the pathway from a professional development programme to student change.

CONSTRUCTS TO BE EXPLORED

We are concerned here with the effect of a set of variables which mediate between the input of participation in the PD programme and the ultimate outcome of pupil cognitive gains. Such mediating variables might act separately or interact with one another to influence the extent to which participation in CASE CPD is translated into students' cognitive growth. The mediating variables we chose to investigate were:

1 Factors within the school management, including:
   a) management commitment
   b) unity of vision
   c) profile of CASE within and outside the school
   d) management's reasons for buying in CASE
   e) CASE co-ordinators' reasons for promoting CASE.
2 The perceived effectiveness of communication about the project within the school science department including both informal modes of communication and formal communication systems within the school related to CASE;
3 The sense of ownership felt by each teacher of the CASE methods;
4 Teachers' attitudes to and familiarity with the theoretical bases of CASE;

In chapter 11 we tie the empirical work reported here into the extensive literature on professional development, but we should here at least point to some justifications for selecting this particular set of variables for investigation. Many authors have highlighted the central and critical role that senior management plays in the effective implementation of educational innovation, perhaps the most recent and comprehensive being Michael Fullan (2001) who shows how, in spite of much fairy dust spread by management gurus, there is actual evidence for the positive effect of managerial coherence (pp 67 et seq.) and for the value of particular leadership styles. The same source (pp 84 et seq.) provides justification for exploring communication as an important variable in implementing school change, and Andy Hargreaves (1994) confirms that

> "Research evidence also suggests that the confidence that comes with collegial sharing and support leads to greater readiness to experiment and take risks, and with it a commitment to continuous improvement among teachers as a recognised part of their professional obligation" p.186

An emphasis on theory comes mostly from our own commitment to the philosophy that one cannot ask teachers to change their practice without sharing with them the underlying rationale, or theory. To attempt to by-pass theory is to treat teachers as technicians rather than as professionals. Technicians learn specific techniques, but are not expected to be flexible in the face of changing circumstances, whilst professionals who have internalised the driving principles can apply them in many contexts. This approach relates to Fullan & Stiegelbauer's (1991) claim for the power of learning-rich organisations in implementing change (p. 331). In a radical educational innovation such as cognitive acceleration, everyone is learning anew: students, teachers, school managers, and PD providers together.

Miles & Huberman (1984) describe the need for a conceptual framework to indicate the supposed relationship between the variables to be assessed. An initial conceptual framework for this study is shown in Figure 6.1 although this simple form fails to show the inevitable interaction which will occur between the mediating variables. The first two of these are school-level variables, while the last two operate predominantly at the level of the individual teacher. Between the school-level and teacher-level mediating variables and the ultimate outcome of student cognitive gain, there is an important intermediate outcome variable, that is the level of use by each teacher of the cognitive acceleration materials and methods. It was important to establish the extent to which teachers were actually using the methods before reading too much into the influence of the mediating variables.

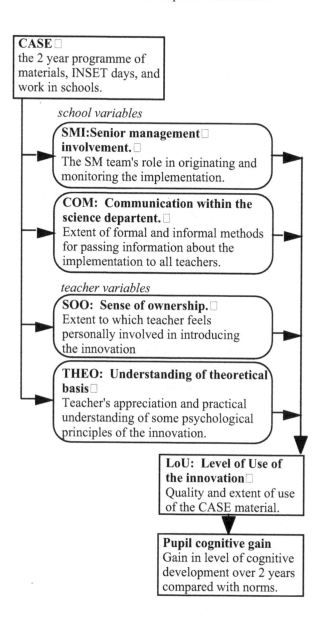

Figure 6.1:A Preliminary Model of Implementation of Professional Development

68 An Implementation Model

PROBES

The four mediating and two outcome variables in this model were probed as follows:

Senior Management Involvement (SMI)

This was assessed by interview with headteachers, heads of science, and CASE co-ordinators using a semi-structured interview schedule. Questions probed each of the elements of this complex variable and answers were followed up as seemed appropriate in order to pursue these purposes. Inevitably, the interviews provided a source of information about other factors which although not included in the original set of hypotheses might also possibly be implicated in the overall effectiveness of the PD, and the extraction of such variables will be described in the second part of this chapter. Figure 6.2 shows an example of such an interview schedule.

Deciding to go for CASE
How was the decision made to buy into CASE INSET?
If you did not make the decision entirely by yourself, who else was involved in making the decision?
Who would you judge to have been the key decision maker?
Motivation
What did you hope your school would gain from CASE? i.e. what was your main motivation for agreeing
    to the expense and time of CASE INSET?
Did you consider that your science department were in particular need of help? If so in what way? OR,
    did you consider that your science department were particularly likely to make good use of a new
    innovation and to follow it seriously?
Introducing
How was the decision to participate in CASE INSET relayed to the teachers involved?
To what extent did all science teachers participate in the introduction of CASE?
Did you feel that there was some persuasion needed? If so, what advantages / incentives were offered?
Do you have an impression of whether the project is being widely accepted or resisted in the school /
    department?
Monitoring
To whom do you delegate responsibility for the routine management of CASE?
What do you do to monitor the progress of CASE?
Do you observe CASE lessons, or discuss progress or results?
Do you expect reports from the department? written or verbal?
Do you make any other checks on progress of the project?
Do you see CASE as something to boast about to parents, governors, OfSTED?
Do you provide reports about CASE to anyone? Parents, Governors, ...
Does the subject of CASE ever come up at Senior management or HOD meetings?
Finale
From a management point of view, has the CASE INSET programme caused any problems or
    difficulties?
How do you see the future of CASE in the school in three years time?

Figure 6.2: Interview Schedule for headteachers

Communication, Sense of Ownership, Theoretical understanding

A 19 item questionnaire was drafted with groups of items designed to probe respondents' perception of effectiveness of communication (COO), their Sense of Ownership (SOO), and the extent to which they had internalised the important underlying theoretical bases of cognitive acceleration. The draft was shared with faculty colleagues who provided written comments and discussion at an internal seminar, and was then piloted with samples of teachers. Item responses were analysed for coherence within each of the constructs and appropriate adjustments made to the wording and to the scoring rules. Figure 6.3 shows a sample of the questions.

SOO

5   Whose idea it was to introduce CASE to your school? (e.g. the head, one teacher, an Authority adviser, a senior management group, don't know, etc.)?

6   Did you participate in any way in the decision to introduce CASE? If so in what way? (e.g. initiator, member of INSET committee, in department discussions, etc.)

COM

9   When other teachers attend CASE INSET days, how do they report back? (tick one or more)
     At a departmental or other meeting
     By a written report or memo
     Informally "over coffee"
     Other ways (please specify)
     There is normally no report back

10   Please make an estimate of the total departmental meeting time over the past two years devoted to discussion of CASE <u>teaching methods</u>. Do not count here either informal chats, or time in meetings spent on the administrative aspects of CASE (such as who is going to do what lesson next, or problems with apparatus. I am asking only for a 'guesstimate'.
              less than 1 hour
              1 to 3 hours
              3+ to 5 hours
              more than 5 hours

THEO

17   Please put one tick next to each of the following statements to show how much you agree or disagree with it. *(4 point scale from strongly agree to strongly disagree):*
     a.   I am only happy if some specific content has been covered in every lesson
     b.   Year 7 pupils are generally capable of reflecting about their own thinking.
     c.   It does not matter if pupils sometimes leave a class a bit confused.
     d.   I can give my pupils a lot of information, if they only listen and make good notes.
     e.   Year 8 pupils can learn to check their own learning successes and weaknesses.
     f.   ....

Figure 6.3: Extract from Questionnaire

For the 'THEO' construct, there are 12 Lickert-type attitude statements with 4 possible choices (strongly agree to strongly disagree). The direction of agreement /disagreement with the CASE theoretical model varied. For example "It does not matter if pupils sometimes leave the class a bit confused" was scored 3 for strongly agree, and 0 for strongly disagree, while "The most effective way of imparting knowledge is for the teacher to talk and the pupils to listen" was scored 0 for strongly agree and 3 for strongly disagree. An internal consistency check (Cronbach alpha with Horst modification for varying facilities) showed that two of the sub-items did not contribute to the same construct as the others, and inspection of their wording showed them to be ambiguous or over-strongly worded (e.g. "In *each* lesson, it is *essential* ..."). A THEO score was computed from the remaining 10 sub-items.

The questionnaire was posted to every teacher in the programme in the years in which this study was conducted (typically 7 to 12 teachers in each of 8 to 12 schools), toward the end of their two-year involvement. A reply-paid envelope was included, and we made follow-up phone calls to heads of science. For those from whom we still did not get responses, we added appropriate questions to the Level of Use interviews (see below). Altogether returns in the order of 75% were obtained.

Level of Use

The Level of Use (LoU) of CASE by each teacher was determined using a Level of Use scale developed by Hall and Loucks (1977). This relies on a tightly structured interview with each teacher which follows a set pattern of probes and subsidiary questions depending on the direction in which the answers are leading. The interview is taped and subsequently analysed according to guidelines provided by the original authors in a comprehensive manual. Figure 6.4 offers a skeleton of the interview schedule, but it does not do justice to the depth of subsidiary questions which are asked at every step. Underlying the ascription of a numerical value to Level of Use is a philosophy which accords the highest level to a teacher who has absorbed the materials and methods of the innovation and then modified them and taken ownership of them to suit her or his particular circumstances. This is a philosophy to which we subscribe. The LoU analysis yields a score on a scale from 0 (is not using the innovation and has no intention of using it) to 7 (is not only using the innovation comprehensively, but has modified it to suit local conditions while retaining the essential main features).

Each LoU interview takes approximately 15 minutes to conduct and requires another 15 minutes for analysis. It is therefore impractical to interview every teacher in a CASE PD cohort, and so we chose samples for interview which ensured a cross section of those directly involved in the INSET days and those who received the professional development programme solely through in-school experiences, as well as including more experienced teachers as well as newly qualified teachers.

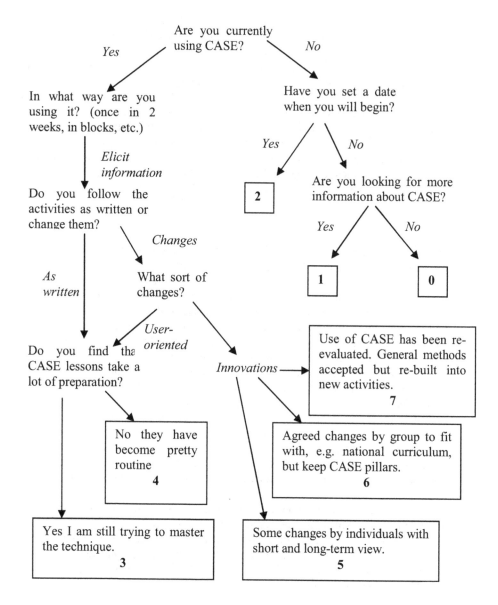

Figure 6.4: Outline of Levels of Use Interview Path. Adapted from Hall & Loucks, (1977)

Cognitive Gains

Cognitive levels are assessed with Piagetian Reasoning Tasks (Shayer & Adey, 1981; Shayer & Wylam, 1978), demonstrated class tasks which yield levels of cognitive development of individuals in a group on a scale from preoperational thinking through to mature formal operations. Schools on the CASE PD programme administer one of these tasks at the beginning and one at the end of the two year CASE intervention, from which cognitive gains can be found for each student, and averaged for classes and for schools. The extensive use of these tasks as quantitative measures of gains made by CASE schools compared with control groups has been described in chapter 5. It should be noted here that there is a direct relationship between cognitive gains made over the two year CASE programme and subsequent academic achievement as measured by grades gained at GCSE taken two and three years after the intervention (Adey & Shayer, 1993).

RESULTS

Two separate studies have been conducted using broadly the same method and instruments, the first in 1993 at the end of the first cohort of CASE PD and the second in 1996 at the end of the fourth cohort. The results to be reported here come predominantly from the first of these studies which involved over 100 teachers from 13 schools. However, in that first study the construct of 'theoretical understanding' had not been properly conceptualised or probed, so for this we need to draw on data from the second study, which involved 88 teachers from 11 schools. In neither study were we successful in collecting complete sets of data.

We have described in chapter 3 the rather hurried initiation of the CASE PD programme. All thirteen schools involved in that first cohort had comprehensive (non-selective) intakes. Nine were within the Greater London area, five were definitely Inner City schools, and one was definitely rural. The extent to which the headteachers and heads of science (HoS) of these schools might be considered 'volunteers' in the CASE INSET programme varied from one in which both headteacher and HoS almost bullied their way into the course in their enthusiasm, to three schools which were informed by their local authority inspectors that they were to participate.

In the sections which follow we will summarise some data-processing methods and relationships which have emerged, and indicate possible significances of these relationships. This chapter will conclude with some general discussion of these results but a fuller account of the implications for professional development in general will not be provided until Part 3, when all of the varied evidence of chapters 5 to 9 have been presented.

Level of Use and Cognitive Gain

From the first sample LoU data was obtained from 40 teachers and cognitive gain data from 35 classes, but both measures together were only available from 18 teachers / classes. The relationship between the variables is illustrated in Figure 6.5. The Spearman Rank correlation between Level of Use and students' cognitive gain was 0.60 (p<.01). Bear in mind that this sample of teachers were all involved in the CASE PD programme and so the range of values of LoU was necessarily restricted. One might suppose that a more representative sample of teachers including some who had never heard of CASE would have shown an even stronger relationship between Level of Use and cognitive gains.

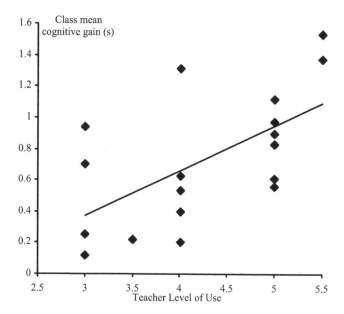

Figure 6.5: Relationship between teachers' Levels of Use of CASE and the cognitive gains of their students

The significance of this result is that it links the cognitive gains – reported at school and cohort level in the last chapter – directly to the extent to which teachers report they are actually using the methods of cognitive acceleration. This is important since it provides evidence for a strong causal link between CA methods and cognitive gains, rather than to more general school-level effects.

Level of Use and Communication

It is natural that amongst the teachers in one school there will be a range of perceptions of the effectiveness of communication about an innovation. What is important is an overall measure from those involved of the extent to which the new methods being introduced are discussed and problems and successes shared within the department, without undue weight being accorded to individuals who may be supposed to be providing information (and so are likely to overestimate the effectiveness of communication) or those at the end of communication chains. For this reason, in order to explore the relationship between perceptions of effectiveness of communication and Levels of Use, we used school mean values. Data for the 13 schools in cohort 1 are shown in table 6.1, and the relationship shown in Figure 6.6.

Table 6.1 School mean LoU and Communication scores

School	1	2	3	4	5	6	7	8	9	10	11	12	13
LoU	3.3	4.3	3.5	3.8	1.8	3.8	2.8	3.5	3.7	4.3	3.8	5.2	4.0
COM	4.1	4.5	4.3	3.7	*	3.6	3.0	4.5	5.5	5.3	3.9	7.3	5.5

*In this school the Senior Management was "too busy" to be interviewed. No questionnaire data was received from this school. Experience in the school suggests that it would have a low LoU score.

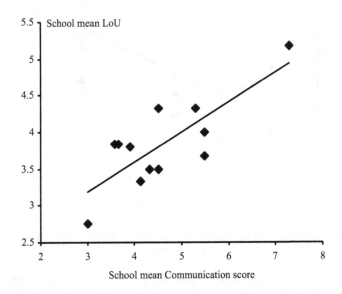

Figure 6.6: Relationship between a school's mean Level of Use score and the mean of its teachers' perceptions of effectiveness of communication.

The Spearman Rank correlation between school means of Level of Use and perception of Communication scores is 0.59. Even without school 12, which appears as something of an outlier, the correlation is 0.49. (Readers who have followed the text carefully since the story of the initiation of the CASE PD programme will not be surprised if we lift the veil of anonymity from school 12 and reveal it as Ninestiles school, in Birmingham. The science teachers in this school would meet together every lunch time in a leisure centre adjoining the school, where the conversation would alternate between social and work-related, including CASE implementation, issues. The warmth of feeling and mutual respect within the department was tangible).

One needs to be careful in drawing conclusions from this result, since the direction of causality is not established. Do teachers who are using the methods more, talk about them more, or does more talking lead to more use? The answer is almost certainly that the two rise together in an iterative process of mutual support such that more talk leads to more use which in turn leads to more talk, in a virtuous circle of development. Even if this is so, the case remains for the value of actually instituting fora where conversations about the innovation can occur, and specific meeting occasions for discussing progress, problems, and successes. Conversely, the observation in a school that little time is being made for such meetings is generally an indication that real Levels of Use of the innovation within a department is low.

Sense of Ownership and Internalisation of Theory

The Lickert scale tapping the extent to which teachers had internalised the underlying theory of cognitive acceleration was added to the questionnaire for the 1996 survey. Data was obtained from 60 teachers (68% of the total sample). It was found that there was a relatively strong relationship between their sense of 'owning' the method and their concordance with the underlying theoretical principles of CASE. This is shown in figure 6.7. The Spearman rank correlation is 0.43.

In a sense this result is just what one would expect, and may fall into that category of psychological experiment which goes to enormous lengths to find out what everyone knew already. On the other hand it does indicate another symbiotic relationship, similar to the last one discussed, where being invited to share the rationale of an innovation and to get inside its theoretical workings gives teachers a sense of professional ownership of the methods. We have repeatedly made the point that real adoption of an innovation includes the ability to modify it, to treat it flexibly, to bring it to bear in new contexts, and the relationship exposed here provides justification for the INSET time devoted (in the particular example of cognitive acceleration) to unpacking the stage-development and schemata ideas of Piaget and the socio-historical model of collective knowledge-building of Vygotsky. The PD literature is replete with claims that change in practice is inextricably linked to change in beliefs about teaching and learning, and we would claim this result as evidence that such change in belief demands attention to theory.

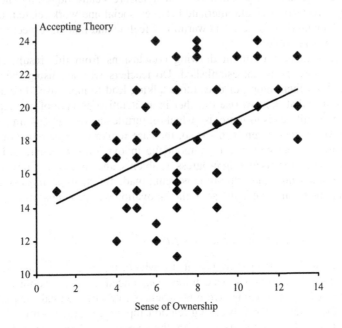

*Figure 6.7: Relationship between teachers' Sense of Ownership and Acceptance of the
Theory of Cognitive Acceleration*

A weaker, but still significant, relationship was found between THEO and COM
(correlation .314, p < .01).

The Senior Management Effect

So far in this chapter we have considered mainly the perceptions and understandings
of the teachers involved in the project. Now attention must be turned to the
influence of the school managers, the headteachers and heads of science. As
described above, the data we have here comes from semi-structured interviews with
each head, HoS, and CASE co-ordinator. The main purposes of these interviews
were to:
• obtain a measure of the 'Senior Management Involvement' factor;
• obtain some independent validation of the 'communication' factor assessed by
  the questionnaire.
In all but one of the thirteen schools of Cohort 1, two interviews were conducted by
the first author in June or July 1993 with the headteacher, or a deputy headteacher
responsible for the science department (or in one case both together) and with the
head of the science department or the CASE co-ordinator if different (and in one

case with both together.) In the exceptional school (numbered 5 in table 6.1 above), only the person who was head of science during the PD programme was interviewed. That school had undergone a radical change of status during the programme, a new Principal[1] having been appointed and all staff having to re-apply for their posts. The new senior management team chose not to reply to repeated telephone messages, faxes, or letters, and when the one deputy principal (who replaced three deputy heads under the old regime) was encountered in the school by chance, she proved to know nothing about CASE or the PD programme.

All 25 interviews were taped and transcribed. Data-reduction and coding was based on methods proposed by Miles & Huberman (1984). Without particular regard to the original purposes of the interviews, all transcripts were scanned for potentially interesting themes which recurred more than once. Naturally many of these were themes which had been prompted by the prepared questions. Each theme of potential interest was made the heading of a column in a matrix. Rows were labelled "headteacher and/or deputy head" and "head of science and/or CASE Co-ordinator" and in each matrix cell were placed direct quotations from the interviews. As this process was applied to the interviews from successive schools, some new themes appeared, and some original themes were modified, and the common matrix altered accordingly. Eventually one matrix was produced for each school, each consisting of twelve columns and two rows (except for that of the exceptional school described above which had only one row). The themes which emerged were:

- Origins:
  Source of original information about CASE
  Who initiated CASE INSET in the school?
  Reason for buying into CASE INSET
  What involvement did the local authority have?
- Communication:
  ... between senior management team (SMT) and head of science (HoS) / CASE co-ordinator (CC)
  ... between HoS/CC and department
- Arrangements for monitoring by SMT
- The profile of CASE outside the school
- Perception of science department
- Other SMT involvement
- Feedback arrangements after INSET days
- Other constraints and problems

A print-out of the first draft of each school matrix was then inspected alongside the original interview transcripts, with three questions in mind:

1 Have all significant statements in the interview been incorporated in the matrix?
2 Are there empty cells in the matrix? Is there any data either already transferred or not which is relevant to this empty cell?

---

[1] As noted previously, the normal English term would be 'head' or 'headteacher'. The use of this American term was a intended by the Authority to signal a completely new management approach.

3   Are the transferred quotations all in the most appropriate cells of the matrix?
A further transfer of data was effected to maximise positive answers to questions 1
& 3, negative to 2. In some cases the answer to question 1 led to "other interesting
dialogue" being added below the matrix.

These second-edition matrices were then given to two colleagues at King's
College London with a request to place the 13 schools in rank order according to a
series of criteria (which we adjusted in the light of early attempts), giving
justifications for their ranking based on internal evidence. These researchers had not
been involved in the PD programme and did not know the identity of the schools.
Agreement between the three judges was estimated using Kendall's coefficient of
concordance, W (Guilford & Fruchter, 1978) and chi-squared. W and chi-squared
values and statistical significance of the agreement for each criterion are shown in
Table 6.2.

Table 6.2: Significance of three judges' agreement on rankings:

Criterion:	1	2	3	4	5a	5b	6a	6b
W:	0.22	0.58	0.80	0.79	0.47	0.59	0.57	0.56
χ^2:	6.48	20.82	19.11	23.74	15.55	21.11	10.29	18.63
p<	.05	.001	.001	.001	.001	.001	0.01	.001

The criteria are shown in Table 6.3
The next stage in the process was to calculate the mean ranking of the three
judges for each school on each criterion and then to look for any systematic
relationships between these mean rankings and the other variables being studied,
that is, Level of Use, Perception of Communication, and Sense of Ownership (using
school mean scores). Table 6.4 shows the Spearman Rank Correlations between
each of these variables expressed as a school mean and the school ranking on each
of the factors extracted from the interviews with heads, HoS, etc.

Table 6.3: Factors and criteria for ranking

Factor name	Description	Typical range
1 Unity of vision	Do the SMT and science Dept., represented by HoS and CC, share a vision vis-à-vis CASE purpose and implementation?	compatibility and co-operation: +5 independent views 0 incompatible views -5
2 Leadership / commitment	Does anyone in a <u>management</u> position (SMT or HoS) have a strong driving commitment to make CASE work in the school?	there is someone determined to make it go: +5 the best is willingness to go along with it: 0 don't-carish or resentful: -5
3 science department management style	How is the project presented to, and run in, the science department?	Authoritarian: 'This is what we're going to do': +5 Consultative: 'This looks good to me, what do you think?': 0 Democratic: 'I'll go along with whatever you-all think': -5
4 Presentation outside SMT and science	To what extent is CASE seen as something that school does and should be proud of?	High profile to parents, governors, and other departments: +5 Make presentations as demanded: 0 Keep it all under wraps, within department, etc.: -5
5a Motivation of SMT	What is seen by SMT as the main purpose of taking on CASE?	Pupil intellectual development (good for itself): +5 Staff development (make them think): + 5 GCSE results: 0 Someone outside said it was good, and it was free: -5
5b Motivation of HoS/CC	as above	as above
6a informal communication	To what extent is CASE discussed informally within the department?	weekly comments, chat in passing, over coffee, etc. +5 Occasionally, comes up because of particular difficulties: 0 Never mentioned: -5
6b formal communication	What systems are there for monitoring reporting on CASE activities within the department?	Regular slot in department meetings, or dedicated meeting; formal feedback required after INSET days: +5 INSET day slotted into department meeting irregularly: 0 none: -5

An Implementation Model

Table 6.4: Spearman Rank Correlations of each criterion against other variables

Criterion	No.	LoU	COM	SOO
1 Unity of vision between SMT and HoS/CC	11	0.54*	0.51*	-0.11
compatible - independent - incompatible				
2 Management commitment to make CASE work	12	0.52*	0.28	-0.13
someone determined - willingness - resentful				
3 Departmental management style	9	0.06	-0.42	-0.08
authoritarian - consultative - democratic				
4 CASE promoted outside school?	11	0.16	0.42	0.14
High profile - as demanded - keep quiet about it				
5a Motivation of SMT	12	-0.23	-0.24	0.37
pupil thinking or staff development - GCSE results - it was free				
5b Motivation of HoS / CC	12	0.35	0.43	0.63*
pupil thinking or staff development - GCSE results - it was free				
6a Informal communication	7	-0.26	0.04	-0.54*
regular chats in passing - occasional - never				
6b Formal communication	11	0.13	0.30	0.70*
Regular slot in meetings - irregular INSET time - none.				

* $p<.05$

'No.' refers to the smallest number of schools for which all three variables have values.

Even with this small sample and the problems encountered with missing data, some pointers to factors which seem to be related to effective PD may be extracted. Firstly, the actual use of CASE appears to be related to (1) the extent to which the senior management team and the responsible science teachers share a vision of the purpose and method of implementation, and (2) the extent to which there is a commitment by a senior figure to the implementation of the method. We can add anecdotally to this, comparing two schools: one, which was ranked very high by all three judges, had a new headteacher and a head of science near retirement who worked together with clear mutual respect and support. The HoS perceived CASE as "the most exciting thing to have happened since the early days of the Nuffield projects". This school was characterised not only by high LoU and effect sizes, but by a striking uniformity of effect across all teachers. By contrast, in another school which had agreed to participate in the project when the local authority agreed to pay all of the costs, the deputy headteacher responsible for INSET actually knew very little about CASE. The project was driven in the school by two young enthusiasts, one a newly qualified teacher and the other a visitor from Australia. Continual nagging from these two for more time and resources to be devoted to CASE did not endear the project to the HoS, who saw instruction in content as more important than the development of thinking. Apart from the two enthusiasts, LoU was low and no post-testing was conducted so there was no measure of cognitive gain. This school was ranked low on criteria 1 and 2 by all three judges.

The extent to which teachers communicate with one another about the CASE innovation seems also to be strongly related to the unity of vision amongst senior management. The contrast of this with the non-relationship between commitment and communication is interesting. It is possible for a school to have a committed individual who drives the innovation through without engendering a feeling amongst the teachers that they discuss it much, but when both science and SMT share the vision, the combination creates an atmosphere in which the innovation is more likely to be a topic of conversation.

The fact that there appears to be no relationship between the amount of communication as reported in the SMT / HoS interviews, and as perceived by the teachers themselves seems to represent a genuine difference of perception between the 'officers' and the 'soldiers'. The former are under the impression (or at least think it right to report so in interview) that information is circulated effectively in the department, but teachers have a very different impression.

Teachers' sense of ownership of the project appears to be independent of management's unity or commitment. Rather, it seems to be related to the type of motivation of the CASE Co-ordinator. This becomes plausible when one looks at the criteria for 'high' motivation ranking, which includes adopting CASE because of its staff development potential. Co-ordinators who place a high value on staff development are most likely to involve teachers extensively in discussion about the project.

The negative correlation with level of informal communication is curious and warrants further investigation, especially when contrasted with the strong positive correlation with formal communication, which would be expected. Added to the comments in the last paragraph, it suggests that category 6a, although amenable to reliable ranking by independent judges, may not be a valid reflection of informal communication practices in a school.

CONCLUSION

Considering all of the relationships discussed in this chapter, what conclusions can be drawn about the proposed implementation model shown in Figure 6.1? As stated when this was introduced, that figure does not indicate inter-relationships between the various mediating factors, and we might now tentatively re-draw that model with extra links to indicate those relationships which appear from the studies reported here to be significant. Figure 6.8 essays such a model.

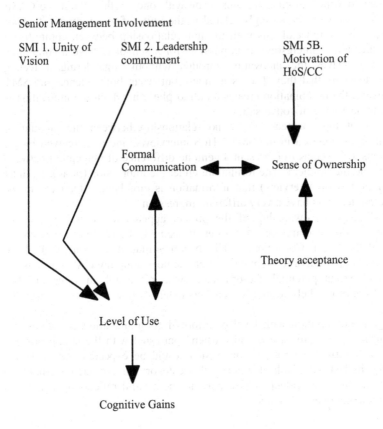

Figure 6.8: A modified model of mediating variables in the implementation of an innovation

Of course, the absence of an arrow does not mean that we deny a relationship exists, only that we have not established one within the studies reported in this chapter. We propose delaying further discussion of the model until we have examined more data, and so we progress to Chapter 7.

7. A LONG-TERM FOLLOW-UP OF SOME CASE SCHOOLS

The data on which this chapter is based was collected by three undergraduate psychology students from Middlesex University who spent one year work experience with us at Kings. I am grateful to Marina Bailey, Joanna Edwards and Nasia Michael, for the energy, enthusiasm, and persistence that they put into this project.

Schools form the nexus of many changes. Cohorts of students pass through, teachers come, are promoted, and leave, administrators retire and are replaced, often the social and economic environment of the school changes, new curriculum demands are imposed and arise from internal review, the buildings themselves decay (rather more rapidly than buildings occupied mainly by adults) and may be renovated or renewed.

Does anything remain constant? Or at least, can a common thread of purpose or practice, of culture or style, be detected running for several years, surviving even radical changes such as a new headteacher or new political imperatives from the outside world? In this chapter we will report an investigation of the long-term impact of the CASE professional development and curriculum programme on one cohort of schools. As described in chapters 3 and 4, CASE PD is a two year programme of INSET days and in-school visits which is relatively expensive to run. For the sake of the schools or authorities who have made the decision to buy into this programme and devoted considerable managerial and teacher time to it, one would hope that the effects were long lasting.

SCOPE OF THE STUDY AND DATA COLLECTION

In this study, we chose to consider as 'long-lasting' any effects that remained visible after a clear year after the end of the two year PD programme, that is three-plus years after the schools had initially 'signed on'. We would claim that this is a rigorous definition of long-term, in the context of the constant flux in schools described above. The work was carried out during the 1997-98 school year, in schools that had participated in the 4th cohort of CASE PD from 1994 to 1996. Of the fourteen original schools, one had withdrawn after just one year, citing financial difficulties. We targeted this study on the remaining thirteen which represented a diverse collection, including three voluntary aided (Catholic) comprehensive schools, two selective schools, one private school, four single-sex schools (three girls, one boys) and eight inner-city schools.

We initially wrote to the head of science of each school, requesting their co-operation in our study, and then followed up with telephone calls to make initial arrangements. We proposed to interview the head of science, the erstwhile CASE co-ordinator (if still there), and some other science teachers. A semi-structured teacher interview schedule was developed by the research team and pre-trialled in one non-participating school. The opening question "Are you still using CASE?" was followed by a branching scheme which probed issues such as reasons for continued or non-use, difficulties encountered, financial problems, conflicts with the national curriculum, students' response, and internal and external support for the innovation. Listening again to the tapes of these interviews some six years later I am struck by the willingness of busy teachers to talk very openly to the researchers. There is a wealth of information here but for this chapter we will focus only on the factors in the schools which seemed to facilitate or to hinder the continued use of CASE after the initial impetus.

The researchers did also devise a questionnaire for year 9 or 10 students – those who were the first to have gone through the CASE intervention – to probe for their memories of and reactions to the 'experiment', but this questionnaire did not yield results sufficiently striking to justify further reporting or analysis here.

All data collection took place between November 1997 and April 1998. We failed to get into three of the schools for reasons summarised below. In each of the remaining ten, two researchers would go to conduct the interviews, which were taped. Here we will present the results as a series of vignettes of each of the schools based on all of the data obtained from that school and which attempt to provide answers to the overall question about the remaining influence of CASE in the school. (To describe these vignettes as 'case studies' would be to promise more detail of 10 schools than can be provided in a single chapter). Each vignette will offer:

- a brief objective description of the school characteristics
- a summary of their level of use of CASE at the time of the data collection
- illustrative quotations from the interviews
- reference to data from 1996 interviews where appropriate
- a brief commentary on factors which seem particularly important in that school.

Initially we placed the thirteen schools into one of four categories: 'No data', 'No CASE use', 'Partial CASE use' and 'Full CASE use'. The schools, all of which have been given pseudonyms, will be discussed under these four categories.

NO DATA OBTAINED

Greenbank School

A mixed comprehensive school in modern buildings in a suburban area with predominantly Anglo students. The head of science had been appointed to his post in 1994 with a specific remit – which he enthusiastically endorsed – of introducing CASE. However, during the two year PD programme, in spite of repeated requests, this school did not find time for the CASE tutors to make more than two of the five coaching visits. On both occasions it seemed that there was strong resistance to CASE from many of the science teachers, and that this was not being challenged by the HoS. When we approached the school for this follow-up work in 1998 the head of science told the researchers that his school had not been properly served by the PD provider who had taken the school's money and not given value. He refused to have anything to do with the 1998 enquiry.

From a 1996 interview with the head of science:

> "... one of the things I needed to do when I came here was to - the department was in a very bad, a very bad state. The head of department had been sacked and was still in the department. ... One of the things that needed, well a number of things that needed to be done but one of them was to establish a culture of teaching philosophy which was effective and to which people could aspire and feel good about. And the thing about your one (CASE) it has statistical verification."

Comment: It is difficult to be sure what was going on here. Observations and the 1996 interviews had indicated that there were some very recalcitrant members of the department, very resistant to any innovation. It seems likely that the HoS was brought in to make changes, but in spite of CASE he had not been successful, and did not want the causes investigated.

South City Estate School

Inner city mixed comprehensive, with students from a very wide range of ethnic origins and linguistic backgrounds. Purpose-built in the 1950s or early '60s, the building is now showing signs of wear and tear.

There was a 'lack of communication' when one of us phoned to cancel an appointment because of illness, the message was not relayed, and the school subsequently refused to allow us to continue.

Comment: This was a great pity, because this school would have been an interesting case. CASE had been introduced by a charismatic young teacher who had fired the headteacher and head of science with enthusiasm. But he had moved rapidly, being promoted to head of science in another school and then soon to acting head in a school on 'special measures'. In 1996 the school had seemed enthusiastic to continue, and it may well be that they were a continuing 'full use' CASE school.

2001 postscript: This school contacted us for a 'renewal' CASE PD programme and at the time of writing we are working in the school on a regular basis.

Bishop Paul School

Inner city Catholic boys comprehensive. Possibly 15% students of Afro-Caribbean ethnicity and a smaller proportion of ethnically Asian students.

In spite of many phone calls and promises to call back, and the geographical ease of access of this school, it never proved possible to arrange a time to visit.

Comment: Reports from this school at the time of implementation show an interesting combination of expressed enthusiasm with actual lack of implementation. They often failed to send anyone to the INSET days, and it was apparent from visits that only one of the teachers, not the HoS, was really interested in CASE and only some of the teachers were using CASE. In 1996 the head of science had said

> I think initially we were happy to give it a go. I mean obviously anything that can improve exam results or what exam results signify, people understand science better. I mean that's what we looked at. I think the response since then ... I don't know how to express it: there's a lot of reaction, I think, to it although it's hard going and heavy work, and a number of staff are not completely convinced of its effectiveness. I mean it does break down our main scheme of things, it is hard to fit in. Some of the material ... with some classes, I mean you wonder why you bothered.

One has to put this down as a probably failure, perhaps because there was never a critical mass of teachers of sufficient seniority with the enthusiasm to ensure a serious attempt at implementation.

NON-USERS

Everton

This is an inner city mixed comprehensive school in an area where voluntary aided (denominational) schools are considered superior, so it has a very low ability level intake.

They are no longer using CASE. The HoS claims that the previous CC (who has left to a university post) did nothing to share the expertise with others, and used the programme primarily for his personal development. We have no data on this school from 1996.

Comment: The HoS's story is credible. Unusually, only one teacher from this school (the CC) ever attended the centre-based INSET days and on our few coaching visits, we only ever saw this person teach. We had some difficulty getting payment from this school, and it transpired that the CC had 'volunteered' the school without clearing the financial implications with the headteacher or HoS.

North Hill

A selective suburban school, mixed. The students are well above average ability and come from a wide ethnic mixture including many Asians, largely middles class.

They are not doing CASE any more, although many of the original staff are still there. They claimed that the main reason was lack of time. This school starts in year 8 and they perceived that they had to cover the whole Key Stage 3 curriculum in two years so there was no time for CASE.

When interviewed in 1996, the head of science and CASE Co-ordinator both expressed great enthusiasm and plans for continuing with CASE. The CC was new to the responsibility, the previous CC having recently been promoted to a new school as HoS in 1996. When he was about to depart, two other teachers claimed to have already established exactly what was needed to continue CASE "... we have committed ourselves to continuing (CASE) from next year". The story in 1998 was very different. Teachers reported that (a) only one person – an enthusiast who was an ex-King's student - had attended the INSET days and he had not passed on the principles, so other teachers were just following the text, and (b) the head of science was notably unenthusiastic about the project.

Comment: This seems to illustrate, amongst other things, the dangers of relying on unsubstantiated interview data. The story obtained by the psychology students from Middlesex University, perceived as independent of King's, was very different from that obtained by interviewers directly from the project. It is also worth remembering that teaching with CASE is hard work, very much more demanding than normal lessons on teachers' effort in the classroom and willingness to put their understanding of the teaching-learning process into practice. In a school with able and well-behaved students, who achieve adequate examination results, the justification for this extra effort may not seem so obvious as in school with greater academic and management problems.

Stadium

An inner city mixed comprehensive school with a very diverse ethnic and home-language mix. Although there are language problems, the intake ability is at about the 35th percentile, so this is not as difficult a population as many of the schools we work with.

They are no longer doing CASE, at least not in any sense as it was introduced and intended. All but one member of the science staff who had participated in the original programme have moved on. The new department knew virtually nothing about CASE, although they expressed curiosity about the test materials and worksheets they had found. When the researchers interviewed teachers here, it stirred up some interest in CASE and I was asked to talk to the department about the possibility of reviving it. One younger member of staff was keen to try, but the new HoS decided that "there was not time".

Comment: This is a clear illustration that the curriculum resides in the knowledge, understanding, and practice of individuals, and not in any physical material (books, worksheets, or technology). A department may participate for two years in an intensive PD program and purchase all the relevant materials, but when the people move on, leaving the materials, the innovation dies. It raises the question for school administrators: can any system be put in place to maximise the chance that professional knowledge is passed on to new members of staff?

PARTIAL USERS

Valley School

This is a rural comprehensive school in a local authority which still maintains selective (Grammar) schools, so this school would have few students in the top 20% of the ability range nationally. It is a relatively small school (4 form entry) with nice modern buildings and well-appointed labs and almost entirely Anglo students.

Some members of the science department are still teaching the CASE activities, but it is not departmental policy and they feel rather rudderless. The original CC, an enthusiastic young teacher who had recently joined the profession, has been promoted to another school, and although the HoS has no particular objections, she sees no need to continue to promote the project which "does not appear on the national curriculum". One teacher said of the CC who had left

> He was really involved, he liked doing all the work. I think everyone's got a bit of responsibility in the department now, it's a bit of a hassle to say to someone 'will you take CASE on' in the way that he did it because he really did go for it".

In 1996 the headteacher had said

> I don't think we set down a set of aims and objectives at the beginning. I was keen to support the enthusiasm of a department. I am happy to support curriculum change, curriculum innovation within the department. … We're not going to drop it because he (the CC) has gone. I think it's now embedded in the department and I think it is now seen as part of the department's approach to the Key Stage 3 curriculum.

Comment: Here we have a headteacher who seems keen to give departments freedom to make their own decisions, but may not be fully aware of the attitudes and forces within the department. The HoS had seemed keen enough in 1996, but without anyone positively driving the CASE programme forward, and with one or two members of the science department somewhat reluctant, there seems not to have been any momentum to keep up the extra work that is required.

Hillingsborough

An inner city girls school, very ethnically diverse. This is a challenging school with a difficult catchment area, which nevertheless has a good reputation for strong and caring leadership.

From 1994, the head of department and one other teacher ('Jane') attended all of the INSET and became extremely enthusiastic about the project. The HoS has since moved to another school as a deputy head (where he has introduced CASE) and Jane has become head of department. However, only one other teacher ('Lilly') remains in the school who was in the original group who were trained by the HoS and Jane. Only Jane and Lilly were interviewed. Both remain extremely enthusiastic about CASE and are teaching it fully, as intended. However, Jane as HoS expressed considerable concerns about the way in which other members of the department viewed CASE and how they were implementing it. Lilly said bluntly that some of them "hate it, from their guts"! Jane was clear that without time to reflect on the underlying theory, it is difficult to persuade reluctant teachers why they should take time out of the regular curriculum for CASE. Three of the current department are temporary teachers and Jane sees no point in sending them for training. Under pressure from the department, she is seriously considering cutting down some of the CASE activities, especially those which the girls find difficult. Both teachers made the point that thinking requires attention, and that CASE could not be done when there are a large proportion of very disruptive students in a class – Lilly referred to only one of her classes being like this.

Comment: There is a classic issue of maintenance here. Without either a structured programme to induct new teachers into the theory and practice of CASE, nor a reasonably stable staff for whom such a programme would be worthwhile, it is virtually impossible to maintain a special teaching programme which makes the demands on pedagogical skills that CASE does. Since this 1998 interview, Jane has left the school and is now a science consultant in a neighbouring borough.

FULL USERS

Dockside

This is an inner city mixed comprehensive boys school in a severely disadvantaged area. Students are of a wide ethnic mix but predominantly working class, and many have English as an additional language. The two teachers involved in the original programme remain enthusiastic implementers, but they admit that it is not as easy to maintain the in-school professional development as it was during the two year PD programme. Nevertheless CASE is built into the timetable – not exactly as initially recommended, but adapted to meet what they perceive as the needs of this school. (Remember that the highest Level of Use on the Hall and Loucks-Horsley scale is 'adapting the innovation while maintaining its essential features'). The head of science reported that out of nine teachers in his department, the young ones were generally enthusiastic about CASE but two older members just followed the activities without really understanding or caring about their nature. Senior Management in the school take a laissez-faire attitude: "If you want to teach CASE, you can" (HoS).

Comment: This is something of a success story in a difficult school, driven by two keen teachers who respect and support each other and who are in a sufficiently senior position to maintain the innovation in spite of real difficulties.

Cathedral

This is a private boys' selective school. Physics, chemistry, and biology are taught separately by different teachers, but CASE is built into the timetable so that all teachers of years 7 and 8 take some of the activities, whatever their specialism. The researchers interviewed five teachers here and encountered a high level of enthusiasm for CASE. The head of science gave credit to the enthusiasm of the CASE coordinator, but it was clearly a whole department initiative. The CC had made a point of taking a different member of staff to accompany her to each of the INSET days so that the sense of ownership was spread. One note of caution was sounded by the CC when she was asked "Is everyone happy?"

> Well that would be a bit of a blanket statement (pause) it would be nice to think so!

This suggests that there may have been some selection of those to be interviewed. This CC also expressed concern about inducting new teachers. This school does not have a high turnover of teachers, but she felt it necessary to devote a fortnightly INSET session to two teachers who had joined the department since the beginning of the PD programme, and wondered how practical it would be to continue this in the long term. On the other hand, although this CC was retiring, she had already identified a successor who had been sent to a CASE convention as part of her induction.

Comment: This is another success story in a school rather different from the majority with which we work. The department is working well together in spite of some reservations from a subject specialist point of view, and there appears to be real forward planning.

St. Agnes

This is an inner city girls' Catholic comprehensive school. This school had chosen to implement CASE across both the science and the mathematics departments, as they perceived that the time required to be taken out of the normal curriculum could be spread across two subject areas. Topics in CASE which looked more mathematical (for instance those to do directly with proportionality and probability) were taught by maths teachers. This was before the CAME programme had been developed. A group of science and mathematics teachers were interviewed together with two interviewers. The original CASE co-ordinator had left the school, there was a new head of science, and a new headteacher, but CASE was still being fully implemented in the two departments, being built into the timetable.

A number of issues emerged from this school. Firstly, there seemed to be a marked difference in perception between the science and maths teachers of the success and viability of CASE. Generally the scientists were very positive about the programme, claiming that the markedly improved national curriculum test results were directly attributable to CASE, and that the girls really enjoyed the lessons (but see below). They agreed that teaching CASE was hard work but saw no point in trying to integrate the CASE lessons with the regular curriculum. Typically, one science teacher said

> I've gained a lot form CASE. I think in a lot of lessons they find they've worked really hard. When I've had support teachers in they've said 'my goodness you got them to work hard'. It really does force them to think and think and think. … Trying to get them thinking well I'm not asking for the answer I'm just asking for a way of how you find out that answer in a way that will give you good evidence and I think CASE is very beneficial for that. It really does get them there.

Mathematics teachers were rather less enthusiastic. They were concerned that it seemed impossible to finish an activity in the 50 minute lessons available to them, that it was too difficult for the less able girls, and that it was impossible for one teacher to manage by themselves. One female maths teacher, who had used CASE only with two lower ability Year 8 groups, was concerned about apparent discontinuity in CASE lessons:

> They never know what the outcome is, if you try to bring them back to it another time, they've forgotten it. They've gone home, they know it was a CASE lesson, but they'd not remember anything … You can refer one maths lesson to another, but with CASE they know it stops. Once this lesson's over, it's finished. The thinking skills haven't but the topic has.

Another maths teacher, male, was concerned about continuing professional development in the department:

> I'm in favour of the philosophy behind it but there's a major handicap, that is training. When I first joined here last year I was just given CASE to get on with it. It was only when I started doing my MA course at King's College when I did the CAME and then I realised that it's deeper than I had thought, there's more to it than teaching.

There was an interesting debate about the extent to which the girls enjoyed the activities. Most science teachers said that they did, but one – more thoughtful perhaps because teachers' perceptions of their students' enjoyment are not especially valid – said

> I think a lot of them didn't enjoy it because they found it challenging, and I think some of these questions (in the questionnaire for students) 'did you enjoy the CASE lessons', I'm sure a lot of people will say they didn't enjoy them very much but I think they got a lot out of them (interjection from maths teacher 'they couldn't remember them!').

Comment: On the positive side it is encouraging to see that the innovation is being fully maintained some years after its initiation, in spite of a number of key changes amongst senior science teachers and senior management. This suggests that it was initially well established with structures which survived changes in personnel. On the other hand, the experiment of asking the mathematics department to share the teaching of CASE seems to have been less successful. I do remember the very first meeting I had in that school in 1994 with the headteacher and heads of science and mathematics. I had the distinct impression that the headteacher's strategy was to use a dynamic science department and the cognitive acceleration project together as a lever into a mathematics department which needed some revitalisation. While it is fair that a mathematics department might resent being instructed to teach activities labelled as 'science' and might also be inexperienced in organising practical work (and so less good at managing time in such activities), I suspect that there is a story here about different management cultures in the two departments in this school.

Riverside Girls Grammar

This is a selective girls' school in a suburban area of the country where all children are still tested at 11+ to determine what type of secondary school they go to. CASE is built firmly into the school's science curriculum, with CASE lessons at specified times throughout years 7 and 8. Parents and school governors are continually given information about the programme and the students are encouraged to see CASE lessons as different form normal science, where they do not bring lab coats or books – "only their pencils and their brains".

The implementation and maintenance are driven by a very enthusiastic CASE co-ordinator who attended all of the INSET days in the 1994-96 PD programme, but enthusiasm for CASE seems neither to be restricted to this individual nor simply to be externally driven by him. It is clear that his dedication to CASE is infectious and also that he makes a considerable effort to introduce new teachers in the department to CASE, taking them through the underlying theory and providing copies of his own notes to supplement the teachers' guide of *Thinking Science*. An NQT reported that he was always available for advice and discussion, but there is no evidence from the interviews of any peer coaching, which raises the possibility (as acknowledged by one of the teachers) that they may not all actually be implementing the methods as envisaged by the originators of cognitive acceleration.

The CC's account of the difficulties and satisfactions of teaching CASE reveal a deep-seated understanding of the principles and methods. He refers to the hard work involved and need to focus on the quality of discussion rather than on the practical work, and believes that while CASE lessons can be disastrous if they go badly, when they go well they are superb!

> To my mind CASE is all about the bursting bubble … that's the moment, that's CASE to me, getting the bubble to burst … 'I've been striving with a problem and suddenly I can see a way through it'. That's the magic … that can happen from the top to the bottom of the (ability) spectrum.

Most of the teachers interviewed believed that CASE had changed their teaching. One claimed to 'teach that way anyway' but also that CASE lessons were good for developing group and independent thinking

> because in normal curriculum teaching you tend to go through quite fast and they don't get a chance to discuss as much as they could.

All of the teachers interviewed, including the head of department who took no year 7 or 8 classes herself, claimed that the thinking skills of the current year 10, the first to have had CASE, were notably improved in spite of the fact that this was not a particularly able year group.

Comment: It would seem that the undoubted success of the implementation of CASE in this school is underpinned by at least two important factors: the continuing enthusiasm and activity of the CASE co-ordinator, and the time and space which is allowed to him by senior management to continue to induct new teachers and support all year 7 and 8 teachers in using the new methods. The headteacher had been enthusiastic about the programme from its inception in the school in 1994. She had ensured that governors and parents were kept informed, mentioned CASE in her annual report and in the school prospectus. In 1996 she had said to me:

> Because we've got able children here we did want to encourage departments to look at encouraging youngsters how to think about problems rather than perhaps always concentrating on content, and we certainly believed that (CASE) could help in that, you know. ... I think particularly as a selective school there tends to be, if I may say so, a traditional approach to teaching which is teacher-led, and really there needed to be more pupil participation and ownership of their own, you know what I mean?

Thus strong support was available for the innovation at all levels in the school.

St. Francis

Thus is a mixed, comprehensive Catholic school in a seaport city environment with many inner-city characteristics.

CASE is still very much a part of the science curriculum for years 7 and 8, but interviews with the head of science and with three other teachers gave the distinct impression while working well together as a group, the department collectively perceived a number of serious difficulties with maintaining CASE teaching. The two main ones were that they saw CASE as being unsuitable for lower ability students, and "impossible" to teach properly in the time available. They felt that they could spare only one 50 minute lesson every two weeks because of the other demands of the content curriculum, and in 50 minutes could not teach CASE as it should be. The head of science said "There's a tight rope there which, frankly, I keep falling off." They also felt that the language of the worksheets was rather difficult.

Listening to the interviews again six years after they were recorded and nine years after that school started on the CA PD programme, my immediate reaction is "Oh those were legitimate complaints then but we have now addressed them through revisions both of the materials and of the PD programme". But one has to ask why this particular school makes such a fuss over these issues which schools with similar or more difficult environments in the same cohort seem to have coped with. There is much nominal enthusiasm expressed by all of the interviewees, but it is seasoned with remarks which suggest it is little more than skin-deep.

> I have to appear to be enthusiastic although I find it very hard; I'm not enthusiastic about it because it's got so many problems with that it doesn't make me enthusiastic but I try not to let that show at all.

There are many positive aspects to implementation here, such as regular meetings at which the HoS discusses progress of the activities, especially for new teachers, although more experienced ones like to go also "because I need help" and there was a clear sense of mutual respect amongst the teachers. Listening carefully to the head of science, however, I started to form a hypothesis that he was, unwittingly, the root of the problem in the department. He was the person who had initiated CASE in the school and had persuaded the headteacher of its value – but almost entirely on the instrumental grounds of promises in gains in GCSE grades. He was an experienced physics teacher who obviously liked children and saw his job as a vocation. In spite of all this we, the PD providers, seem to have failed to make any significant change in his fundamental beliefs about the nature of intelligence or the nature of teaching and learning:

> What CASE does I think is help youngsters maximise their thinking ability. It does take them on as far as they are able to go, but not every soldier is going to be a general and CASE won't turn youngsters with a reading age of 7 and a CAT score of 70 into erudite thinking scientists. That's got to be tackled on a much broader front.

"Taking them as far as they are able to go" suggests the notion of a fixed ceiling.

> The lower ability ones need more encouragement to get going, they need to have their confidence boosted that they can do it. And they're often surprised when they find that they can "Oh I see, oh is that it? Oh I can do that" that's rewarding and yet frustrating. You knew that 45 minutes ago, all that time, so you think 'if I could start now' but of course you can't, that's it.

Implicit here, and in many other places in the interview, is a 'delivery' view of the curriculum. Why don't they listen to me at the beginning when I tell them they can do it? Because learning is not like that! Similarly he expressed the view that if the worksheets were better written, it would "free the teachers from the burden of explaining". Here is the will-o'-the-wisp of the crystal clear worksheet which any student simply needs to read to know what to do, and how to think.

Comment: I suggest that these fundamental beliefs made it difficult for the head of science to challenge the reservations of colleagues who had not been on the PD programme. Perhaps, were they to attend more recent versions of our PD programme they would have had an opportunity both to hear our reassurances that 'finishing' an activity was of no importance, that you could even skip the practical work and give out data to think about, and also have heard more of the experiences from other schools who met the same concerns in various ways. At the same time, I must accept the despite our often-expressed realisation that fundamental changes in practice will not come about without significant changes in beliefs, in this case we seem to have had only limited success.

DISCUSSION

In our search for factors which have an impact on the effectiveness of professional development, the data from this long-term follow-up study reinforces some of the suggestions which emerged in chapter 6 and highlights some additional factors which seem to be important. Senior management commitment, unity of vision between the headteacher and head of science, and the motivation of the head of science and / or CASE co-ordinator are all reinforced as important factors, as is the need for continuing formal communication within the department about the innovation. There is also some evidence for the role of theory in encouraging attitude change. Additionally, it seems clear from this data that a necessary condition for successful maintenance of the innovation after the end of the PD programme is the establishment within the school of structures which provide a constant framework as teachers come and go. The maintenance of an innovation requires strategic planning, and the establishment of systems for both maintenance and renewal. For example, it seems important that CASE lessons are specifically written into the timetable, and not just left to teachers' whims about when they will teach them. Likewise, the schools that are successfully maintaining CASE two years after the end of the PD programme have a system for inducting new teachers into the background theory which also contextualises the intervention in the particular conditions of that school. They have support systems for teachers new to CASE.

Perhaps most obviously, this study shows the centrality of personal factors in maintaining an innovation in a school. Where an individual trained had not passed on any of the ideas and had then left, nothing remained. Where there had been a large scale exodus of those trained, nothing remained. It is teachers, individual people, who are the subject of the professional development programme, not some abstract entity such 'the department' or 'the school'. If the people move, that expertise moves with them – although it may well be transferred to another school.

8. TEACHERS IN THE SCHOOL CONTEXT

This chapter presents Nicki Landau's work. The 'I' here is Nicki.

THE RESEARCH SPACE AND CONTEXT

The aim of the research reported in this chapter was to disentangle some of the characteristics that seemed favourable to professional development, that is, which caused a positive response to a particular PD programme. The study starts from a deceptively simple hypothesis: the effectiveness of a professional development programme will be mediated by (a) individual teachers' stances and personalities, and (b) the school environment in which the teacher finds him/herself. Very crudely four extreme situations can be envisaged as a 2 x 2 matrix:

		School Ethos	
Teacher Stance		Unsupportive	Supportive
	Positive		
	Negative		

It is not difficult to predict the fate of an innovation in an unsupportive school with negatively-inclined teachers. Nor is it anything for an innovator to be especially proud of if his/her programme is enthusiastically adopted in a supportive school whose teachers have a generally positive attitude to new ideas. What is far more interesting than either of these (admittedly stereotypical) situations is the complexity of the real world where school environments cannot be crudely characterised as supportive or unsupportive, where teachers of one subject work within a department which forms its own ecological niche nested within the school, and where teachers even in one department demonstrate a multi-dimensional range of personalities and stances towards new ideas.

This study focuses on a small group of schools and some of their teachers who participated in either the Cognitive Acceleration through Mathematics Education (CAME) or Cognitive Acceleration through science Education (CASE) Professional Development programmes that ran over the two school years from July 1998 to July 2000. Before describing the schools, some attention should be paid to the methodology of this sort of study.

97

METHODOLOGICAL ISSUES

My aim was to chart how teachers were interacting with the PD programme's materials and methods on a personal level and in their classrooms, through personal contact and growing familiarity with them and their working environments. Using some observation of teachers' lessons and by listening to their reflections and ideas about their practice throughout the two year programme I planned to gather detailed data that signalled patterns of behaviour associated with effective change. My interpretations of key teachers' words from informal staff-room chats and from more formal interviews and observations of their actions in and out of class, all contributed to a picture of their beliefs and perceptions of events. While empathy was important if I was to see the CA project's invasion into their working lives through each teacher's eyes, it was also crucial to objectify and clarify through triangulation, examining their stories alongside other key teachers' stories from the same department and looking for consistencies between word and deed. Hence, I peppered the conviction

> that the way people talk about their lives is of significance, that the language they use
> and the connections they make reveal the world that they see and in which they act"
> (Gilligan 1993 p. 2)

with some questions relating to the causes of their actions and interpretations within their working context.

Projects such as CASE and CAME become uniquely interpreted within each department and re-interpreted by each participating member, so a dual view was essential. My attention was directed by the complexity of the context in which teachers were operating and the interplay between the individual as a professional, the department as an organisation and the project's outward reachings towards its audience. All the data collection was done responsively, except for field notes taken from documents, in an effort to capture the richness of each teacher's individual story.

The idea that a teacher's individuality, as a professional, is pivotal to their pedagogic practices being amenable to changes during a PD programme was rooted in my background as a secondary school mathematics teacher, having undergone a plethora of PD courses some of which continued to develop and inform my classroom practice. Empathy with teachers in this study came easily as the realities of classroom, departmental and school life had dominated my working life for so long. Seeing and experiencing the CA induction process through key teachers' eyes was a crucial aspect of the study. However, the image I presented to heads of department (HoD) and to teachers was a cause for concern with regards to how it might affect their responses to me. I was studying at the same institution as the CA providers, and this might lead teachers to assume that I had some authority in relation to the project. Such an assumption would be wholly false, since I myself had never even undergone the CA PD during my teaching career, let alone become part of the training team. It was therefore necessary to de-construct myself as far as possible, as expert or judge or advisor, in the eyes of the staff of the four schools studied. My ignorance about teaching CA lessons helped, because it provided a means by which I could un-expertise myself in teachers' eyes. To maintain a non-threatening position it was necessary to distance myself from the CA provider and stress the teacher's expertise in their school, their department, their students and the realities of the project. My lack of experience was helpful and my stance was honesty rather than collusion, since my one trial at teaching a *Thinking Mathematics* (TM) (Adhami, Johnson, & Shayer, 1998) lesson revealed the difficulty of teaching for cognitive acceleration, my own inadequacy and the need for consistent practice. Like my key teachers, I was unable to enact or describe or visualise exactly what an effective CA lesson really was. In this way, I was struggling with the teachers I was observing. Nevertheless my focuses in classroom observations necessarily grew out of my understanding of the theory and my studies of the lesson guidelines. I identified some of the key aspects of CA lessons and used them to focus my teacher and pupil observations from a theoretical rather than a practical grounding.

The challenge of my task was to discover through interpretative social research methods the course of a PD project's two year induction in four schools, aspects of success as defined by some change in teachers' classroom behaviour, and their possible causes.

DATA COLLECTION

The strategy was to focus on a range of teachers' experiences and journeys through the initial two year CAME or CASE project within contrasting schools. Table 7.1 summarises the opportunities taken for data collection and the material generated.

Table 7.1: Summary of data collection events and products.

Date	Event	Outcome
1st school year		
9/98	Negotiate with HoD, teachers, for teachers to follow	Field notes
10/98, 1/99	Observe provider's INSET days	Field notes
10-12/98	Lesson observations: 1 CA, 1 non-CA lesson per teacher	Field notes
	Interviews with teachers	Transcripts
4-6/99	Attend school meetings, read inspection reports[1]	Field notes
	Observe providers' INSET days	Field notes
	Lesson observations: 1 CA, 1 non-CA lesson per teacher	Field notes
	Interviews with teachers	Transcripts
2nd school year		

(there were now only 10 of the original 15 teachers still in the schools, and one of the four departments left the project in January 2000)

---

[1] In England the Office for Standards in Education (OfSTED), a government body, conducts in-depth formal inspections of schools at least every five years. Their reports are published.

1-6/00	Maintain contact with remaining teachers and departments as circumstances allow	Field notes
	Lesson observations: 1 CA, 1 non-CA lesson per teacher	Field notes
	Interviews with teachers	Transcripts
	Attend providers' INSET days	Field notes

3rd school year (the PD programme has now formally been completed)

| 9-12/00 | Maintain contact with the non-participating second year school | Field notes |
| | Attempt to follow progress of other schools into the third year. | Field notes |

A few more notes are in order on the nature of each type of data collection event.

Lesson Observations

In an age when the purpose of classroom observations are frequently either inspection by an external body (OfSTED) or appraisal by a senior member of staff, it was important to create a clearly positive and non-judgmental position. With this aim in mind, teachers' comments on the quality or pertinence of the project's lesson plans, ideas and materials, as well as their pupils' responses, were always encouraged, especially during debriefings. The key teachers' expertise was manifestly valued in discussions as was their knowledge of the teaching groups, the curriculum within their department, their subject matter, and peculiarities of their school.

Within the classroom, while I attempted to record in note form as much as possible, I also tried to provide some positive support for the teachers in their classroom activities, whether handing out work sheets, books or equipment, eliciting pupils' explanations about their work or directing their attention towards the task, or assisting in any clearing away. The intention was to be helpful in some way while recording as much of the classroom activities as possible. Observation schedules and tick-box lists were rejected because of their constraints and limitations. As I transcribed the field notes which included time spans of particular activities, examples of types of questions and responses and levels of distraction or absorption all jotted on a reporters pad, I would recall the lesson in my mind and how it was to be a part of it, for both pupils and teacher. I was able to obtain a 'feel' for each key teacher's style of classroom techniques and detect any changes in it during their department's initial instigation of the CA teaching programme. The lesson observations aimed to answer questions such as: How did their classroom practice change? If it changed for CA lessons, did the change carry over into other lessons? What might have helped to motivate any change? What factors might help facilitate PD? Although the teachers' beliefs about learning and teaching went unclassified by tests or specific questioning in most cases their classroom behaviour frequently implied their views. Since during the CA lessons a new set of beliefs and behaviours were meant to be adopted, the non-CA lessons were likely to prove more revealing about genuine teacher change.

Teachers' tension levels during observation varied from barely noticeable and with me being warmly welcomed into the spirit of the lesson, to open admissions of nerves but greater familiarity through ongoing contact eased concerns as indicated by their increased nonchalance towards electing lessons for observation. A short debriefing session allowed teachers to comment in some way upon the lesson and some one-to-one on-task discussion with pupils was common to all lesson observations.

Interviews

The aim of the interviews was to gain qualitative data about the key teachers and their perspectives on both their departmental experience and personal professional journey during the two years of CA PD. They were usually arranged towards the end of each school data collection phase, so that we were more relaxed in each others company and lesson observations could be referred to, in the hope of fruitful outcomes from a potentially threatening situation. Interviews were usually held in teachers' class bases but occasionally in staff working areas during free periods. Each interview session was recorded and generally lasted between 20 and 30 minutes, occasionally longer. Every interview was fully transcribed and then checked by the interviewee who was encouraged to signify any unquotable sections, check for accuracy of meaning, possibly insert words that had been difficult to transcribe, and add any further pertinent comments if they wished. Most transcriptions were read through but only a few were annotated by teachers. Both the openness and opportunity of some control offered by this strategy helped to create a more trusting relationship over time and may have reduced some subjectivity. When offered subsequent interview transcriptions, teachers indicated a growing trust and confidence by specific comments, such as "I'm sure it's all fine".

Interviewing techniques were focused (Cohen & Manion, 1994) and aimed to glean the maximum information possible about the key teachers and their experiences of the CA project at their schools, in a flexible and responsive manner. Three interview schedules were sequentially prepared providing a framework for questioning and prompting that helped ensure against omissions. However, considerable flexibility was demanded to achieve a recording of each teacher's personal story, as their circumstances differed and changed. Interviews became increasingly open-ended and unstructured, allowing teachers' talk to flow into areas they wished to raise or felt pertinent.

Field notes

These were made on all other occasions and on documents from which I could glean information about the teachers, the schools, or the PD programme. Examples include

- meetings in the schools (including some training sessions)
- staff discussions
- documents
- the CA INSET days
- telephone contacts with CA school co-ordinators

SAMPLE

The Schools

Four schools were chosen from among those enrolled in the providers' 1998 to 2000 CAME and CASE cohorts for their:

1 geographical convenience, within a 20 mile radius of the metropolis;
2 potential for access and acceptance, usually because a member of staff had past links with the provider's institution, and
3 range of contrasting features.

The two CAME schools, which I have called Citysite Comprehensive and Greenbelt High, exhibited the obvious contrasts of a falling role, inner-city community school and an outer suburb, commuter belt, popular and partially selective institution, respectively. In the classrooms and corridors of Greenbelt, orderly student behaviour seemed regulated by commonly held expectations which were clearly absent at Citysite, where weak management compounded the challenges facing teachers in coping with students' behaviour. In both mathematics departments the HoD seemed very positively disposed towards the innovation. The other teachers, some of whom had been at the schools for many years, included a wide range of attitudes towards the project and its inception. In both schools, although the HoDs worked closely with at least one other chosen colleague, no one was appointed as official co-ordinator of the project and their interpretation of their role as leaders during the CA initiation differed. At Citysite, there were many immediate demands on the HoD's time, because the department was understaffed and their published performance table GCSE results had gone down recently. Little extra in-school time could be allocated to CAME and when additional meetings were attempted, contending with an undermined and incomplete staff resulted in frustrating further attempts. In contrast, Greenbelt's mathematics department consisted of a full complement of mostly longstanding mathematics teachers and the HoD was able to pursue his enthusiasm for the project. Early in the programme he negotiated with senior management and with his department for after school meetings and this was ultimately appreciated as the whole department benefited from his efforts organizing all of the teaching materials as well as his vision of collaboration.

The two CASE schools, dubbed Crossroads High School for Boys and Woodview Comprehensive, were popular suburban institutions and their CASE co-ordinators both had past associations with the provider. Their contrasts were subtler (except that Crossroads had an all male student population) than those exhibited by the two CAME schools, relating to their institutional ethos and department members' explicit attitudes towards the innovation. From the start of the project, despite a keen co-ordinator, it was evident that there was some overt hostility towards CASE within Crossroad's science department. Woodview was part of a Teacher Training Consortium offering staff free access to Diploma Courses and consequently the number of science teachers involved in their own professional development or in the training of others, or both, was unusually high and created an ethos of valuing teacher learning within the department. Crossroad's HoD, having delegated the position of CASE co-ordinator unofficially, was pro-active in his collaboration with her but reluctant to impose extra demands on all his staff. At Woodview, the responsibility of CASE was officially added to those of Key Stage 3 co-ordinator and the HoD facilitated some requests for extra in-school time specifically for CASE discussion or INSET.

All four schools sent at least one department representative to almost all of the provider's INSET days in the first year, indicating the positive attitude held by all the faculty heads at the outset. Despite their different interpretations of their departmental role during the CA PD, the four HoDs were both supportive of, and involved in, the project from its inception. Woodview and Citysite both appeared more open to new developments from educational research, one due to its learning ethos and the other because of a willingness to try anything that might help compensate for its students' disadvantages and help improve their results. This was evidenced by their involvement in a number of initiatives. By comparison Greenbelt and Crossroads felt far more traditional in approach, both having been former Grammar schools[1] where teachers frequently take on the role of authoritative expert.

The Teachers

Within each school I planned to work closely with four teachers, but in Crossroads it proved possible to identify only three, so the total number of teachers to be followed from September 1998 was 15. By starting with four teachers in each school it seemed likely that at least two from each would continue into the second year of the PD programme. The plan was to aim for a gender mix and range of

---

[1] Up to the late 1960s selection to all secondary schools in England was on the basis of test taken at 11+ years. Grammar schools took approximately the top 20% of ability children based on this test. After that most Education Authorities changed to a comprehensive system with secondary schools accepting children of all abilities.

teaching experience as well as hoping to include the teachers who seemed closest to, and most distant from, the CA PD induction process. The rationale was to capture a spread across each department's hierarchy, so that issues relating to teachers' positions within faculties, pertaining to the new developments during the course of the innovation, could be addressed. Since HoDs formed a special group they were excluded from becoming key informers, although field notes about the department included their activities. On the other hand all of the co-ordinators were included (whether they were officially appointed or not), together with another teacher of similar experience or responsibility and one or two other members of staff.

These 15 teachers were involved in both the first and second phase of data collection spanning the first year of the project. However, because of staff movement, promotion, time tabling, agreeable access at Citysite, and Crossroads dropping the innovation during the first term of the second year, it became necessary to operate more opportunely in the second year. The third phase, planned for the second year, consisted of ten teachers of the original fifteen, only six of whom were then teaching CA lessons. One of the six teachers, whose story could provide the continuity from the start of the project into a second year, I followed to a new school. Ideally, these key teachers, would have taught the new lessons to at least one Y7 group in the first year and continued through the programme with Y8 groups in the second year, but some schools' time tabling was subject to constraints relating to staff movement, expertise and shortfalls. Hence, only two of the key teachers were definitely continuing with the more advanced material for Y8 classes in the second year.

Analysis

In preparing the case studies I needed to be able to draw from all of the sources: material from documents, meetings, informal discussions, telephone conversations, field notes from the provider's INSETs, interviews with that school's key teachers, and first hand observations. For each of the case study teachers, I prepared a large scrap book into which I lightly pasted bits of transcripts, lesson observations, field notes, and other data, grouped according to particular themes which seemed to be emerging as the analysis proceeded to build a firm evidence base. It was important during the analysis process not to close too quickly but to allow themes and theories continually to arise from the data, to develop and to assume greater or lesser prominence over the whole period of data collection.

THE CASE STUDIES

Space will only allow us to offer three of the 13 case studies here. We judged that it would be more revealing, and illustrative of the methods and of the results, to offer three case studies in full rather than try to summarise many. They are selected to illustrate in some detail an important sample of issues that are germane to the theme of this book. We will not comment on these case studies here but in the analysis of part 3 we will draw on these examples as well as on other examples from this study not presented here.

Rita

Rita was well experienced, having taught mathematics at Greenbelt for 11 years. She had already been second in department for several years when I met her, having returned to teaching part-time after her children's early years and she was gradually accruing more professional responsibility. Her past experience, outside Greenbelt, included teaching at a girls grammar school and some private coaching. Although no one was officially appointed as co-ordinator of the CAME project, because of Rita's position, she was closely involved in the departmental induction process from the start.

Rita attended almost all of the provider's INSET sessions in the first year, accompanied by her head of department (HoD) and the third post holder in the department. Together, they shared the responsibility of supporting their colleagues by cascading the INSET demonstrations of each Thinking Maths (TM) lesson in turn and leading the reflective discussions on previously taught lessons. When necessary these sessions took place during extra after school meetings that the department's staff had agreed would form an essential part of the CAME induction process. This had the effect of maintaining a similar pace of progress through the TM lessons for all the mathematics teaching staff, with Rita and her two (key) colleagues usually leading each trial. This pro-active team of three also decided to organise TM lesson observations across the department. Teachers were grouped into threes and Rita was the first to be observed and then observe another colleague in her group, leading by example. In the Spring term, the core co-ordinating group organised an in-school training session, led by a member of the provider's institution. During this half day session, Rita was involved as a classroom observer and a participant in the lengthy and detailed debriefing that followed.

Rita had an air of organised efficiency and calm authority about her. Teaching appeared to fit into her present life style with considerable ease. She never seemed rushed or unprepared or fazed in any way, taking the pace and demands of school life in her stride. This may have been partly due to her stage in life, having independent children and living alone, she could afford to be generous in her allocation of school focused time. On my first visit to Greenbelt, when I attended a department meeting to be introduced to the staff, she was very helpful and informative. Staying behind with the HoD, she explained the school's in-take and setting procedures, decoded the timetable and was the first to accept being a participant in my research. Within the department a considerable percentage of her teaching was to A level and upper school examination classes, implying her attained status as a mathematics teacher and her post conferred her central position in all the faculties business. All the department's staff seemed to enjoy good working relationships and Rita was clearly a key member of their team. In the staff room, although she tended to occupy the same seat and bring her own packed lunch everyday, she was sociable towards everyone around her.

I first observed Rita teaching a non-CAME lesson towards the end of the Autumn term, the occasion of our fifth meeting. Because the department's staff had such a conscientious approach towards the introduction of CAME and Rita was involved in both trialling and demonstrating TM lessons, she had already taught five at least once. Prior to the pupil's arrival, Rita showed me the exercise booklet used by the department saying: "We're doing number machines from year seven booklet". Most of the lesson was taken up by Rita circulating and helping pupils as they worked quietly through the specified exercises on number machines. In the first five minutes she introduced the topic and demonstrated the prescribed method with pupils supplying the answers of two examples. Rita interrupted their working three times at approximately 15 minute intervals to read out sets of answers, which they were encouraged to mark clearly and honestly, and coaxed to get on and complete the exercise.

This lesson conformed well to Rita's description of her teaching style. During our first interview she had explained her preference for a quiet classroom in which her pupils could "sit and watch me teach on the blackboard" and work on their own. Using the word traditional, she elaborated further by explaining her own inability to "do maths with noise around me" and that "group work is something alien to me".

For the first CAME lesson to be observed, Rita selected TM 7, a well structured lesson in which pupils generate algebraic relationships with two variables leading to their graphical representation, although she had taught the first five CAME lessons sequentially. The frequency of public pupil talk and more open questioning had substantially increased in this lesson. However, the open questions were sometimes followed by more directed or even closed questions that were clearly seeking particular answers. This sometimes led pupils to recognisable guessing strategies. The board work was used for some sharing of pupils' ideas, as well as a tool for demonstration. However, like the first lesson observed, pupils worked individually on the notesheets and there was a detectable emphasis placed on right answers, rather than on contribution. In the non-CAME lesson students were chastised if they were sharing ideas and "getting off the point and chatty", during TM 7 pupils were told that they could leave out filling in the table at the bottom of notesheet two, because "it depends if you can get the answer right or not". In this classroom, following and listening to Rita's demonstrations, instructions and answers was as highly valued as quiet individual endeavour, since pupils were repeatedly told to wait for her explanations before starting on notesheets. So despite the periods of lesson talk increasing by about four times and more open questions, pupils' answers were still constrained and mostly channelled to producing correct answers. Correct answers, rather than pupils' ideas, were seemingly the aim of public pupil talk.

Before teaching this lesson, Rita had demonstrated at least one of the TM lessons to all her colleagues in the mathematics department, observed the remaining cascaded demonstrations and taught the first five TMs, some more than once. It may have been significant that she delayed teaching TM 6, since the content appeared more removed from text book mathematics being about directions, lengths and angle directions, than TM 7 which she chose for my observation. She had previously described how she felt that TM 4, a practical lesson which includes pupils measuring each other, "would have been chaos" without the assistance of a support teacher.

During the second phase of observations and interviews I watched Rita teach another two lessons in the Summer term. The first, a non-CAME lesson, was about converting Centigrade to Fahrenheit and vice versa using a conversion graph. There were strong similarities to this lesson and the first non-CAME lesson observed, although the board demonstration time period was doubled, being 10 minutes on this occasion.. Despite some of Rita's questions being open, like "What do I do?" and "What can I do?", her responses to pupils' answers effectively made them into closed questions. When a pupil suggested: "A conversion, I don't know the formula" and another: "Use a scale" to the first and second questions respectively, Rita either repeated the question another way or seemed to ignore them. In this way, it became clear that on each occasion she was actually looking for a particular answer. The major part of the lesson was spent with pupils working individually through conversion exercises from the same booklet as in the first observation and Rita intermittently reading out the answers. Ten minutes before the end of the lesson she instructed the students to complete up to exercise G for homework and gave them 10 mental arithmetic questions from a national curriculum test paper. The pupils self marked their answers as Rita called them out and then asked for a show of hands to indicate who got them all correct, then one or two wrong before packing up to leave.

The second CAME lesson that I observed, later the same term, was TM 12 about functions and their graphical representation. In this lesson, the three plenary sessions accounted for almost half the lesson time, open questioning had increased and some probing, follow up questions appeared. However, Rita also slipped into a teacher telling mode and closed or directed questioning as often as she used the more open and exploratory style of teaching. Although this lesson signalled some degree of altered classroom behaviour, the implicit differential values of teacher's voice and pupil's voice were still detectable, along with the importance of pupils working individually and right answers. In this classroom, there was little opportunity for right answers to be arrived at through group discussion, building on different student's ideas or misconceptions. Right answers still had to be known or calculated individually. Wrong answers tended to be dismissed or ignored rather than probed or used as an opportunity to invite other pupils to comment. Thus, despite the lesson appearing superficially different in pedagogic style from those seen earlier in the year, it still failed to provide pupils with any opportunities for real discussion or debate based on their own understandings.

There are two aspects to Rita's story that make it significant. The first is how comparatively little change occurred considering her central position in the department during the first year of CAME's induction and the facilitating aspects of her context. Because of the HoD's energy and conviction towards the CAME induction at Greenbelt, and his good relations with both his department's and senior management's personnel, he had created a very positive and nurturing environment for the potential encouragement of Cognitive Acceleration Professional Development (CA PD). He had negotiated extra after school meeting times for, and with, the mathematics department's staff, and with management, for three people to attend INSET training and free cover class time for mutual observations and attending the in-school training debriefing. Additionally, he had stocked a resources cupboard with fully prepared class sets of equipment for each TM lesson to be taught in the first year of for all his mathematics teachers' ease and convenience.

The second is how her classroom practice modified during TM lessons, indicating that real altered pedagogy is quite subtly distinctive from a constant style of teaching that adapts for particular lessons while remaining essentially unchanged. Regarding this second point, it could be that this was an essential first phase of Rita's changing style of teaching. Unfortunately, since she left to teach abroad, we will never know what developments might have occurred in a second year. Nevertheless, taking her context into account Rita's story indicates that for some teachers, however facilitating the departmental context, their style of teaching may have undergone some degree of ossification over time, causing something of a block to change.

At the time of our second interview Rita perceived herself to be teaching "everything else (i.e. non-CAME lessons) very much the way I've always done it", implying an awareness that her classroom style remained essentially unchanged.

Peter

When I met Peter, he was a young teacher in his mid twenties, with three years professional experience behind him. Since starting at Woodview as a Newly Qualified Teacher (NQT), he had acquired two points of responsibility, one pastoral assisting the head of Y7, and the other departmental for the KS3 science curriculum. It was the latter responsibility that was extended to absorb an official CASE project co-ordinator within the department. This worked particularly well in the first year, because Peter was already both familiar with and interested in CASE, as a result of his past association with the provider's institution from his PGCE course.

In the first year, as co-ordinator, Peter was responsible for organising the necessary equipment, preparing the workcards and worksheets, supporting his colleagues and enlisting the support and co-operation of the technicians. As he attended all the provider's INSET days with another member of staff, he was able to delegate some of these activities, as well as collaborate over decisions. He helped to develop a timetable of TS lessons for the department's staff to aid support and collaboration among his peers, and to create a CASE time slot within the regular department meetings agenda for introducing and discussing some of the new material. Peter moved to a new school in the second year, Parkvale, which meant he was neither able to benefit from all the extra work entailed in setting up CASE within Woodview's science curriculum, nor to take his TS lessons into Y8. Parkvale, as a newly established school, provided him with both the opportunity of setting up a science department from scratch, and the potential of becoming its HoD as the school grew. This move also resulted in Peter repeating TS lessons and teaching mathematics, including TM lessons, to mixed ability Y7 groups. He became a key figure in the school, as his expertise of CA teaching techniques was relied on, because of his past CASE training and experience as a co-ordinator, in both the science and mathematics departments.

Peter was outgoing, personable and very friendly, enjoying positive relationships with all members of the school community. His positive energy and ease of mixing with people was demonstrated in both Woodview and Parkvale, where he became a partner in mutually beneficial lunch arrangements, among other things. He seemed to establish good working relationships in Woodview with both the HoD and his peers, some of whom shared the in-school CASE INSET burden. One colleague was enlisted to trial each TS lesson, another to lead a debriefing during a department meeting, and the in-school INSET session was organised collaboratively. Likewise, in Parkvale, he worked constructively with both the Deputy headteacher and his mathematics department peer, on the science and mathematics curriculum, and CA lessons. He collaborated with the Deputy headteacher to develop at least one CASE-style activity and used his experience, as CASE co-ordinator the previous year, to help support his mathematics colleague as she attempted to familiarise herself with CAME and learnt to deliver TM lessons effectively. This was all confirmed by field notes based on observations and conversations, as well as recorded interviews, in which different groups were indicated by a high frequency of 'we' in various contexts, rather than 'I'. Peter's interviews revealed a relatively low, We to I ratio, of one to two.

I first observed Peter teaching a seemingly typical science lesson about the effects, in experiments and on pH scales, of different strengths and concentrations in acids. He used some closed questioning to recall and bridge from past lessons to introduce the topic, followed by a demonstration which led into the students' experiments. He included some copying from the board, instructions for their writing up, and after clearing away, a short class summing up session, heavily teacher directed, with mostly closed questioning with Peter emphasising the main points. This seemed to contradict some of his description of a typical lesson when he had stressed the importance he placed on pupils' talk, discussion and action. This was a teacher directed lesson in which Peter had done most of the talking. 'Public' pupil talk had been confined to mostly short responses to closed questions, addressed to Peter, at the start and end of the lesson for approximately five minutes in total; the rest of the time they were passive listeners. Class participation and activity was confined to their experiment which included some informal discussion with a peer. Teacher interaction was limited, since the plenary sessions were dominated by short pupil responses to closed questioning accompanied by teacher talk, and multiple student interaction and discussion was negligible.

Before this lesson, Peter had taught TS 1 three times, prior to the department ability banding Y7's for science. During his first interview, by which time he had also taught TS 2, he described how "I thought very much more about questioning and them (his pupils) thinking why" in TS lessons. This was validated on the TS 2 lesson observation when open questioning and probing, that invited whole class participation, became a significant feature. This second lesson observation, of TS 2, contained far more pupil talk and included a new requirement, for students to listen to each other in order to agree, disagree or contribute further, under Peter's direction and invitation. His questions were no longer dominated by 'what' and 'when', but included 'how' and 'where', along with requests for further examples and demands for agreement and disagreement. The students' interaction was no longer confined to the teacher, a partner and equipment, but also included their working group and other class members during the final discussion; and the total plenary sessions time had increased by at least 10 minutes.

During the second phase of observations in the third term of the project's inception, the style of teaching witnessed in the TS 2 lesson seemed to become a characteristic of Peter's normal science lessons. His non-TS lesson felt like an adaptation of a TS lesson. The number of open ended questions had increased, but more significantly, the variety of questioning and directed dialogue had grown richer. Now, 'why' questions were introduced along with more probing, that required the same or another student to elaborate on answers already volunteered, as well as an increase in checking for agreement or disagreement, and demands for bridging into the students' outside world. While teaching non-TS lessons these features remained constant and plenary sessions could occupy up to half of the total lesson period. Additionally, there was evidence that Peter had adapted and extended TS 7 to explore ratios more fully. He did this by causing his pupils to 'make a map' of the laboratory with the benches representing land masses, and also, requesting pupils to bring a scaled down model of something to a science lesson for them to calculate the approximate scale that had been used in its creation. Thus, Peter managed to alter his teaching substantially during the first year, adapting and developing TS lessons by the Summer term.

Initially, in preparing himself to teach the new TS lessons, Peter said that he liked to talk through the activity's procedural notes with colleagues. He felt that this helped him to identify different aspects and possible practical difficulties of the lesson within the context of the classroom. Although he felt that TS 2 "was chaotic" because of the student's difficulties with the equipment, he enjoyed the novelty of the new lessons. His attitude towards teaching them seemed positive, despite admitting they were difficult. He liked the increased pupil talk, classroom noise and activity, the focus on reasoning, and he recalled how well TS 1 had gone on his third attempt. His frustrations were dominated by timetabling constraints and organisational obstacles. As co-ordinator, he was constantly having to think about other peoples' next lesson, because the timetable meant he was preparing to teach one TS while facilitating his colleagues to teach the next; he found this quite stressful. He agitatedly described how, "I was frantically cutting tubes, literally an hour before someone's lesson". The gathering together of the newly required equipment and supporting technicians, so that they could support the department's effort, absorbed more time and energy than Peter had anticipated. Another early source of his frustration was the delay in getting CASE started in the department, a result of negotiations with the provider and then teaching delays caused by the re-sorting and streaming of the Y7 in-take. In the Summer term, another cause of his frustration was the department's rate of progress through the TS lessons. Although he conceded that everyone was "trying to keep it going", despite the extra time needed for reading through the notes and preparation. He also talked about gathering and organising the appropriate new apparatus as causing delays and how school activities, such as end of year examinations, reduced the department's teaching time.

However, Peter's frustrations seemed overshadowed by what appeared to be very positive feelings about teaching 'the CASE way' towards the end of the first year, spearheaded by learning to deliver the TS lessons. He claimed to "enjoy the lessons more when I take a CASE approach", when acknowledging that CASE had changed his teaching style. Peter felt he was becoming relatively confident in his delivery and that they were running more smoothly, because he had "more of an idea of what I'm looking for and identifying signs in the class with the lessons being of a different sort of style and concentrating more on thinking than doing". He felt he could identify the signals, indicating when pupils were thinking, meeting the conflicts and trying to work through them. He was becoming less worried by the time constraints of a lesson period and "focused on what the real purpose of it (the TS lesson) is". As he taught an increasing number of TS lessons, he felt that he could piece together the thrust of the lesson after a quick read of the teacher's guide and while actually teaching it, "because you never know quite where a lesson is going to go". He no longer used the procedure details prescriptively, perceiving himself to have gained the freedom to allow both his pupils and himself to alter the course of lessons to facilitate the groups' optimum understanding of problem, rather than follow a rigid plan.

In his classroom practice, Peter saw himself as questioning more "as opposed to teach or dictate what's going on. I try and explore something with the class and get their ideas and what they are thinking about it ... before the questions were probably less open ended". He believed that the desire to find out the reasons behind his students' answers had altered his questioning to "why isn't it anything else, or why do we know, or how can we be sure". He described how he felt his response to wrong answers had altered; "before I would have probably said ... nearly but not quite, has anybody else got an idea, now I'm more likely to say why is that then ... and talk through it, and they can realise ... I've (the student) got this wrong, which is important that they (the students) realise why something is wrong as well as why something's right". In this way it was clear, Peter was conscious of how his classroom practice had altered, which can be verified by the observations. During an informal discussion before leaving Woodview, Peter revealed his belief that CASE helped teachers to teach 'the right way' to get pupils thinking and learning effectively, with its focus on reasoning rather than doing. He felt that the TS lessons took at least 70 minutes and since Woodview's periods were only 60 minutes long, he had been extending them into another lesson so that the crucial fourth and fifth pillars could be fully explored. His consciousness and awareness of the importance of working through all five pillars during a lesson were emphasised by the content of the in-school INSET session which he directed. In this session the department's staff were arranged into small groups to write lesson plans focusing on how the five pillars fitted into them, according to all the interview data of the other teachers.

During Peter's second year, in Parkvale, his major responsibility was to plan and establish the new science department, including curriculum, equipment and texts. He attended none of the second year CASE INSET days at the provider's institution, due to the new school's non-enrolment in the official professional development programme. However, a positive desire to implement and incorporate both CASE and CAME within the curriculum meant a reliance on the printed materials and whatever staff expertise and collaboration he could call on. Peter shared the science teaching of the six Y7 mixed ability tutor groups with the deputy headteacher, who was both a science specialist and CASE trained, and his mathematics department counterpart, who was neither. He also taught his own tutor group mathematics, including TM lessons. This resulted in some mutual reliance between Peter and his mathematics department colleague as they familiarised themselves with the additional Y7 subject curriculum. In learning to teach CAME and guiding his mathematics counterpart, Peter had to rely on his knowledge of CASE. In this task his experience as a co-ordinator, the previous year, and his own professional development seemed crucial.

Peter's lessons continued to feature the newly developed characteristics from the previous year in both TS and TM lessons, but a little less so in the non-TS lesson. This was accounted for by his admission that he had got out of the habit of CASE-style teaching at the start of the second year, as a result of becoming distanced from the official two year PD induction programme. He believed teaching CAME had helped him to re-focus his thinking again, regarding his classroom practice. In his view, the TM lesson plans made the theoretical purpose behind each phase during the teaching period more explicit and dealing with numbers was more straight forward and less distracting than experimental results. This helped both pupil and teacher to focus on thinking rather than getting side-tracked by the frequent inaccuracies and inconsistencies of experimental data in the classroom. Peter continued in his creativity with regard to classroom activities, helping to develop a CASE-style activity in which pupils made 'cars' from used plastic bottles, straws, elastic bands etc., that could be propelled forward for a 'race'. According to Peter, the aim was to encourage the students to think out why and how some 'cars' did or did not work, or achieve higher speeds, and for them to make links with the friction and forces topics from the science curriculum.

Discussing his teaching, he talked about "questioning what they're (the pupils) doing, try to ... cause conflict in their thinking, as they're doing an activity ... in order to test understanding" to help his students resolve and understand what, why and how they know something. He stressed the importance of questioning during pupils' experiments to help focus their thinking while they were engaged in the activity. He also felt that consistently adopting the TS language had provided students with "something concrete to hold on to" and that "the idea of thinking and explaining follows on more ... naturally" from CASE terminology. He felt his students responded differently in CASE lessons "because there's also bridging and applying it to other situations ... there's no 'I've finished now' to it ... they know that when they have finished, what they're going to do is then start applying their knowledge". Although the idea that his students were more engaged, talking better and thinking more were repeated, this latter quote conflicts with his earlier assessment about TS lessons and student response when he indicated a finality about each lesson. "It's almost as if there's something to crack by the end of the lesson and when we've cracked it that's it we've done that ... we can just go".

Towards the end of the second year, having worked collaboratively with his colleague on the science curriculum and mutually supported his mathematics counterpart, his assessment of the CA programmes had developed. His perceptions about his students' responses seemed linked to his feelings about what caused him to alter his classroom practice and his ideas about how both students and teachers learned. He talked about seeing the benefits and value in CASE, as signified by watching students unravelling problems in the classroom, or returning with thought-through solutions after leaving puzzled. He now seemed to believe that "unless you've reacted with a problem and come to some ideas about it yourself, you won't learn anything", stressing the need for his pupils to solve a problem for themselves in order to learn effectively. Students "have to do it for themselves and that can be facilitated by me, or by a group, or by anybody else", "somebody else might put in the final ... piece of the jigsaw for them, but that will only piece together what they already thought, already. It's no good just having something explained".

Reflecting, over the two years, on his experience of teaching CA Peter said "it feels quite foreign in the beginning and then ... either because you see a value in it, or just because you've done it so much, it tends to become a bit of a habit". The reason he gave for changing classroom practice in TS and TM lessons was that it was essential "there's a need to change teaching style for CASE lessons", defining the necessary change as "the way you interact with them (the pupils), questioning and things like that". So it would appear that Peter affected his professional development, initially though his struggles and attempts to teach TS lessons according to the prescribed practice. Then, it was his observations of the positive effects of his identifiable altered classroom practices on his pupils, that helped to fuel the further and more far reaching changes in his teaching generally and possibly his beliefs about teaching and learning. The crucial factors in Peter's story of successful professional development seemed to be his willingness to persevere with the guiding principles of CA lessons, helped by his tolerance of some classroom disorder and his frustrations, and his ability to diagnose the differences in his teaching and recognise the subsequent effects on his pupils responses. Even when the 'habit' of CASE-style teaching left him, he was able to recapture his past success through the same fruitful cycle of perseverance, identification and recognition that led towards the creative re-invention of his classroom practice.

Beth

Beth was highly qualified and had three years teaching experience when I met her. She was in her early thirties, leading the busy life style that is typically associated with parenting a young family alongside full time employment. She had joined Crossroads Comprehensive the preceding September having previously taught in a similar school, within a different local authority. Although she was effectively responsible for the CASE project in the school, she remained a regular teacher without any formal point of responsibility or time allocation for the extra obligations that implementation entailed. Her unofficial status had come about because of her keen interest in CASE, as a result of her past association with the provider's institution during her PGCE course, and the lack of spare funds for a further appointment in the science department.

As CASE co-ordinator, Beth became responsible for organising all the new equipment, enlisting the support and co-operation of the technicians and trying to support her colleagues. Like other official co-ordinators she prepared class sets of worksheets, laminated work cards and, with the technicians, ordered or adapted and arranged the apparatus for each of the TS lessons. In the first year, Beth attended all the provider's INSET sessions, almost exclusively alone. She procured various prominent notice boards in the science block, one along a busy corridor for displaying TS materials and students' work and another in the staff's resource room for recording the TS lessons taught by each teacher, evaluation sheets and other CASE information of interest. Beth and the head of department (HoD) trialled and discussed each TS lesson ahead of their colleagues, sometimes recommending an alteration to the suggested procedure for example proposing the experimentation with tubes in TS3 to be done as a demonstration. To satisfy a complaint about the time consuming nature of CASE lesson preparation, Beth drew up a "crib sheet" of what teachers and pupils needed to do and ask at each stage, during each TS lesson, based on the lesson's guidelines. In this way she made positive efforts to encourage and assist her peers during their early attempts to teach the new lessons, also arranging for a trainer from the provider's institution to come into Crossroads for a practical classroom support and debriefing session.

During the last term of the innovation's first year, Beth began to consider her future prospects in the teaching profession. In early June, she confided that she had been offered a job in another borough, but felt that the second year of CASE afforded interesting challenges for her at Crossroads. Then, towards the end of the month, she unsuccessfully applied for an internal pastoral post. She seemed to take this disappointment philosophically and continued to work positively on the CASE project within the department. She remained responsible for all the necessities of teaching the new lessons and began working on specific timetable slots for the TS lessons of Y7 and Y8 in the following year. This was a complex task because the old Y7 groups would be differently mixed in Y8, as a result of streaming, and some teachers had completed fewer TS lessons than others. However, she persevered in her attempts to overcome this, by pairing teaching groups appropriately, in a determined effort to ensure greater departmental consistency in the progress through CASE lessons during the second year of the project.

Unfortunately, in the second year, much of Beth's planning was wasted as teaching groups were mixed by different criteria and in the Spring term the HoD decided to stop their involvement in the CASE project. He reasoned that the department had completed the time span and used up the financial allowance allocated to the programme. At this stage, Beth revealed that she was considering leaving teaching to concentrate on a business venture which was already partially set up. She described the school environment as disheartening, with teachers begrudging any extra effort because of the school's financial deficit and impending redundancies. During the Summer term Beth gave notice that she would be leaving teaching at the end of the academic year. She described the major factors as disillusion because of school's failure to adequately reward effort, and child minding difficulties.

Beth had a positive, friendly and open personality and quickly revealed her domestic, as well as school, situation. She greeted people with a smile, was lively, open to challenges that utilised her initiative, resilient and business-like. Beth liked to get along with her colleagues and regularly frequented both the department and school staff rooms. (Some science teachers used only one of these areas exclusively). She worked hard on behalf of the HoD when he decided to enrol the department in the CASE programme, including him in her activities and negotiations. She also responded positively to her colleagues' difficulties with the extra lesson preparation. Although she was frequently frustrated from raising the profile of the CASE project within the department by the HoD's preference for not using meeting and INSET time for discussions, she sought out ways to compensate by using notice boards and writing an evaluative report. Beth seemed to accept and worked within her lack of official authority in good spirit and with concern for both her pupils and colleagues in the science department. Her interviews revealed a 1:6 'We to I' ratio. This indicates the degree to which a particular key teacher identifies with her or his departmental colleagues during the inception of the CASE project.

I first observed Beth teaching a lesson about categories of living things. She began with a plenary session lasting 10 minutes, which focused on the general characteristics of living things by asking some open questions to encourage her students to provide specific features. Using the board and some of her pupils' ideas for headings, she provided the class with summary notes to copy into their books saying "if you're asked in a SATs exam you'll be able to write something". The pupils then worked individually on a work sheet and in small groups sorting cards of living things into different categories. Beth linked this classification activity to the way Museums of science and Natural History group displays. The lesson finished with clearing up, praise and homework instructions. This conformed to Beth's description of a typical lesson being teacher led, asking students questions, making sure they had all the information, as in the note copying, and doing most of the talking herself. While she liked to make content interesting by giving anecdotes, her focus was giving out the information necessary for tests. In her opening plenary session pupil responses were short and the little probing that occurred was directed at an individual pupil's response, so although she encouraged them to listen to each other, they could have perceived it as unnecessary.

Before this first observation Beth had taught the first three TS lessons and at the time of her first interview, two weeks later, five. In preparation for teaching TS lessons Beth read through each plan making notes of: "what I should do and what they (the pupils) should do". This included the types of questions she should ask the students at each stage of the lesson and it was these notes that she passed on to her colleagues when they were ready. This careful translation from the suggested lesson procedure details to her own practical list of what she had to do and question her pupils about at each stage proved an effective method by which Beth seemed to manage the necessary classroom innovation. The second and third observations were both TS lessons in which plenary sessions were interspersed throughout and the total time taken by them had tripled. In these plenary periods, Beth started to include quieter pupils by asking them for additional information when another student had raised a point. The variety of her open questioning had increased from the more straight forward: "why is ...?" to: "why do we ...?", "why not ...?" and "what if ...?". These more probing questions were even more frequent in the third observation, which also included Beth asking students if they agreed with what another member of the class had said and why. By this third observation it had become more crucial for pupils to listen to each other and as they described their ideas and methods, Beth's role became that of a confirmer and discussion manager, rather than a teller or leader. Students' public talk had become as significant and almost as frequent as Beth's.

In the second phase of observations, towards the end of the first year of the project, it was apparent that Beth's normal science lessons had more similarities with TS lessons than they had prior to the innovation, as indicated by the first observation. Although copying notes from the board was still a significant feature of her teaching, the questioning was exclusively open and accompanied by frequent probing. Plenary sessions were interspersed throughout the lesson and accounted for the same proportion of time in both types of science lessons observed. The requirement for students to listen to each other was maintained through her direction of public pupil talk being an inclusive exercise of joint participation with pupils contributing towards combined explanations and elaborations.

During the first interview Beth said that she felt her regular classroom approach was starting to change with Y7. A new priority had invaded her teaching, that of "getting out their ideas and thought processes and getting each child to contribute". She was able to identify how her teaching had altered: "I am asking them more questions and recapping more ... getting them involved, saying what did we do, why did we do that ...". Although initially Beth was concerned about teaching TS lessons to her mixed ability group because of the significant number of special needs pupils, at the time of the first interview she already noted that some of her pupils who usually lagged behind were "starting to blossom". She felt her students were more engaged and that their verbal participation and response indicated improved levels of confidence and logical thought processes. However, with higher year groups her focus seemed to remain dominated by the content as she stressed the importance of: "getting that module finished with". By the second interview, towards the end of the project's first year, Beth felt her approach towards topics had altered with all her year groups and was convinced that her students' thinking ability had progressed as a result of TS lessons. She consciously and consistently made time for, and coaxed, pupil discussion by following answers with: "do you agree with that, anyone got any other ideas". Beth perceived this type of questioning as crucial for all her pupils to experience a topic: "then collectively you get the full picture".

Despite Beth's continued enthusiasm and commitment to the CASE project within the department she was struggling to enthuse her colleagues. She was attending all the provider's INSET alone, went to the CA convention alone and although she organised a trainer to visit the school, she was the major beneficiary because it coincided with an OfSTED inspection week. Thus, she struggled to include her colleagues in the learning, sharing and adapting process implicit in a comprehensive departmental adoption of a significant classroom innovation. The HoD had attended the provider's first INSET and he joined Beth for the debriefing session given by the visiting trainer. He also trialled the TS lesson with Beth and they discussed the best way for their colleagues to teach them, but he failed to grant her requests for departmental in-school INSET time. Due to some past personnel tensions in the department Beth thought the HoD was very sensitive to "causing too many waves" and was trying to bring CASE in "through the back door". This resulted in CASE taking a low profile among the general departmental concerns. TS lessons went undiscussed at meetings and science teachers had little opportunity to ask advice openly, make suggestions, contribute evaluations or simply compare groups, lessons and effects. Beth and the HoD informally collaborated and tried to provide support to the Y7 science teachers through 'silent' methods. Avoiding public talk meant that Beth's efforts were channelled into producing written sheets on simplified CASE lesson preparation and evaluation sheets. She also wrote and distributed a department CASE evaluation, which was more like an information news letter about CASE in the department, for the benefit of those teachers who were not involved in the initial teaching of it. Beth was concerned about the lack of consistency in the pace of progress through the lessons of the Y7 science teachers and with the lack of collaboration generally. She found it "difficult to get things done or across" because, despite having the responsibility, she lacked the authority. It was her lack of authority that frustrated her ability to realise the vision she developed of the role of CASE co-ordinator, especially after attending the CASE convention. Nevertheless, she sustained the extra workload and continued to devise ways in which she could effect an improvement in the project's progress during the second year, planning complex timetables and Y8 group divisions.

In the second year of the CASE innovation, I observed Beth teaching two lessons during the early part of the third term. TS teaching in the school had ceased at the beginning of the spring term. The first lesson was a library research session for her pupils, their task being to find out about and prepare a two minute presentation on Volta and one other scientist of their choice (Ampere, Ohm and Watt were suggested and some pupils 'discovered' Edison and Faraday), consisting of an illustrated poster and talk on their lives and contributions. This was clearly an untypical lesson with students working in small groups negotiating, planning and sharing the research and preparation work, supported by the available books, with only minimal reference to Beth. In the second lesson, the first half hour was taken up by the students' presentations. Although self consciously given, they were impressive and lasted approximately three minutes each. The rest of the lesson was devoted to understanding something about batteries. The pupils tested the designated equipment, completed a work sheet and copied up some notes but, significantly, the activities were interspersed with class questioning that accounted for the same proportion of time as in the TS lessons. Beth's questions were all open, there was some probing and less telling than was in evidence during the first observation of the previous year.

During Beth's final interview there were references to two CASE type lessons that she had developed: "living?" and "mud".

KEY FEATURES

These case studies provide rich accounts of the process of implementing a teaching innovation into normal schools. Teachers will empathise with many of the realities portrayed here and PD providers will recognise the successes and frustrations involved in the process of introducing change in schools through an approach which relies on the strength and authority of one or two key players. All of the stories illustrate the fact that real change is a slow process. Initial attempts to employ new teaching methods are inevitably faltering and inadequate but continuing input from INSET sessions and supportive coaching do lead to demonstrable changes in practice which eventually become deep-seated in the 'natural' pedagogy of the teacher. Further, they show that such changes diffuse from the special CA lessons to the normal teaching practice of the individual across all of his or her lessons.

But the studies reveal also the detail of how the process can fail, because it is embodied in individuals. Where the individual is in a strong leadership position, and is in or can induce a collegial supportive atmosphere, then the innovation catches on and becomes "what we do". But where, as in Beth's case, the first line recipient of the PD is neither given recognition nor support in introducing new methods, she becomes isolated and the department shields itself from the innovation. The result is worse than simple non-adoption; it is the disillusion and loss from the profession of someone who, in only mildly different circumstances, might have become a teacher-leader who could have had a long-term multiplier effect on the quality of education in a school, in a local authority, or wider.

The features we will pick up from these case studies for further discussion in the concluding chapters include the roles of collegiality, of leadership, and of the nature of the PD programme in contributing to the effectiveness of any PD .

9. MAKING THE PROCESS SYSTEMIC: EVALUATION OF AN AUTHORITY PROGRAMME

This chapter is the work of Gwen and John Hewitt. The 'we' in this chapter is the Hewitts.

The CA@KS1 Cognitive Acceleration programme (*Let's Think!*) was outlined in chapter 4. It will be recalled that it consists of about 30 activities which incorporate the five pillars of Cognitive Acceleration (CA) and which are based on the schemata of concrete operations. 1998-99 was a developmental year in which activities were designed and piloted in selected schools. In 1999-00 the training was further refined and was evaluated by monitoring changes in the children's cognition, as described in chapter 5, and in subsequent years new groups of teachers and schools have been incorporated into the programme. The first three years of this programme (1998-2001) involved teachers exclusively from the London Borough of Hammersmith and Fulham (who funded the original research and development) but it is now being widely extended throughout the United Kingdom. All of the teachers in Hammersmith and Fulham (H&F) were provided with intensive Professional Development as described in chapter 4. However, such a PD programme was relatively expensive to deliver and problematic in the context of a borough with very rapid turnover of young teachers. For example, in 2000-2001 only eight out of fifteen teachers who had trained in the previous year remained in the borough, and three of those moved with their class from Year 1 to Year 2. Therefore, in order to maintain the status quo, ten new teachers had to be trained for schools which were already involved in the project, before new schools could be included in the programme.

One of the aims of the third year of the project (2000-01) was to increase expertise in the borough by enabling experienced CA teachers to become teacher-tutors. In 2001-2002 these teacher-tutors were to become involved in preparing a new cohort of teacher-tutors. It was reasoned that this ongoing cycle of PD should maintain the methods of cognitive acceleration indefinitely, eventually needing only minimal and occasional 'expert' input. Cognitive acceleration would then become 'systemic', part of the education system of the borough, and would be wholly owned and controlled within the borough by the inspectorate and the headteachers.

OUR RESEARCH

We joined the CA@KS1(*Let's Think!*) project as researchers in January 2001 and our brief was to evaluate the 2000-2001 systemic professional development programme. Our interpretation of this brief was that we should provide evidence to help determine if such a programme was both viable and sustainable by describing the process and identifying areas for change and development. Our research focused on the perceptions of the teachers, the teacher-tutors, the headteachers and the local education authority link inspectors. Questions were grouped into two broad categories. The first related to the impact of teaching *Let's Think!* in the classroom, for example, is the professional development involving teacher-tutors perceived as enhancing the teachers' pedagogy, the children's learning skills and the children's cognitive development? What factors influence the impact of *Let's Think!* in the classroom?

The second related to the systemic professional development programme, including how it was perceived by the teachers and the teacher-tutors and by their headteachers and their link inspectors? How did the teachers view the contribution of the teacher-tutors to their professional development? How did the teacher-tutors perceive their own tutor professional development programme? Were there any problems with teacher-tutors working with teachers? Additional questions included: How did *Let's Think!* and the systemic professional development programme affect the schools? Could the professional development 'package' of centre-based work and school-based work be reduced further in order to reduce the costs?

Our data has come mainly from semi-structured audio-taped interviews which we carried out with all of the teachers and the teacher-tutors, firstly in January and again in June. We also interviewed all of the headteachers and the relevant link inspectors in June. We attended all of the centre-based professional development days for teachers and teacher-tutors and we visited schools to observe *Let's Think!* sessions. We had access to the evaluation questionnaires ('opinionaires') from individual professional development sessions from both 2000-2001 and 1999-2000 and to the lengthy open-ended questionnaire administered at the end of the year for the three years 1998-2001. We also looked at the written feedback given to teachers by teacher-tutors. All interviews were transcribed and themes were first identified by each of us individually and then collated jointly. Some quotations have been edited in order to make them easier to follow. We used a mixture of qualitative and quantitative methods. The quantitative methods and analyses were presented in chapter 5 and here we focus on the qualitative data.

The PD programme for *Let's Think!* was outlined in chapter 4. More specifically, in 2000-2001 the teachers' PD programme consisted of:
- 6 centre-based PD days (3 in term one; 3 in term two)
- 6 school visits to each teacher (2-3 demonstrations mainly in term one; 3-4 observations & feedback mainly in term two).
- 4 peer observation visits to other *Let's Think!* teachers (2 in term one; 2 in term two)
- 1 optional twilight session in a local school in term three

Six teachers, who had all completed the *Let's Think!* PD in June 2000, went on to follow a tutors' professional development programme and to act as teacher-tutors in 2000-2001. Previously, both the in-school work and centre-based PD had been delivered by borough and university staff. These teacher-tutors were fully involved in the professional development work in schools and also contributed to the centre-based sessions. We could find no other reports of schemes of this kind in primary schools in this country. The teacher-tutors attended a professional development programme designed to equip them to work with teachers both in school and on centre based days which included:

- 5 centre-based days, (3 in term one; 2 in term two);
- organising and delivering a session at the national CA convention;
- being accompanied on their initial visits to schools by borough or university staff;

During the year the teacher-tutors were funded for:

- 10 days of visits to teachers' schools (terms one & two) where they first demonstrated *Let's Think!* activities in the classroom and later observed teachers' *Let's Think!* sessions and gave feedback
- 2 days participation in the teachers' professional development days where they simulated *Let's Think!* activities, led discussions and delivered individual sessions on CA pedagogy and organisation.

The teacher-tutors also organised and attended one twilight session for teachers in term three.

PARTICIPANT TEACHERS AND SCHOOLS

Seventeen new Y1 teachers joined the programme in September 2000 and the teaching experience profile of this group shows that they were relatively inexperienced. The six teacher-tutors had more experience but were still in the early stages of their teaching career. The profile of both groups is shown in table 9.1, and of CA schools and teachers in Hammersmith and Fulham over three years in table 9.2.

Table 9.1: Profile of Teachers and Tutors in the 2000-2001 Professional Development Programme; Years of Experience in September 2000

	NQTs	1	2-4	5-8	8 +	Mean (SD)
No. of teachers:	6	5	3	2	1*	2.7 (2.2)*
No. of teacher-tutors:	-	-	3	2	1	5.3 (3.5)

* This teacher had been teaching for 28 years. She is excluded from the Mean and SD

All of the teacher-tutors had taught for at least two years in H&F, and the first cohort of teacher-tutors had taught in H&F for an average of 4 years.

Table 9.2: Profile of the Involvement of H&F Schools and Teachers in CASE

	1999-2000	2000-2001	2001-2002
CA schools:	10	16	22
CA teachers in schools:	14	26	38
Experienced CA teachers:	-	8	16
New CA teachers starting PD:	-	18	22

THE IMPACT OF CA IN THE CLASSROOM

In chapter 5 we presented some data on the effect of *Let's Think!* on measures of children's cognitive growth. Here we are concerned with the teachers' perceptions of the effects of CA on their pupils. Analyses of written evaluations and interviews showed that 60-70% of teachers thought that they were very or quite successful in implementing *Let's Think!*, 19% had some problems and 11% were unable to implement it fully. Observations, written evaluations and information from the interviews of teachers, headteachers and link inspectors all point to a range of changes in the learning skills of children and the pedagogy of teachers involved in the programme. Overall, teachers, headteachers and link inspectors were very positive about *Let's Think!*. For example, from teachers:

> ....and they love the problem solving, and they get a great buzz out of it when I question them. This is not just in *Let's Think!* but in their Maths and English, and even when we are doing PE, in fact in every subject.

> I think that the children's whole thinking has developed over this year, not just their problem solving. Now I put that down to *Let's Think!*, I really do. I certainly think that its got them thinking just that little bit further, thinking "why, why did this happen". It's got them to ask the questions.

> At the beginning they just thought that it was a game but now they have realised that it is a challenge and that they have to think. This is what I value. It's not whether they are right or wrong, it's their opinion that is important and the fact that we have introduced words like 'disagree' and 'agree'. You wouldn't normally teach this to 5 year olds, because it is quite a difficult concept.

Some headteachers' comments:

> Well, it makes the children much more independent of thought, but at the same time able to work co-operatively. They are giving consideration to other people's points of view, not just listening quietly to them, but actually taking on board what they are saying and evaluating it. This is something that you don't often see in children, particularly ones that young. Often, they are quite egocentric, but this is making them consider other people's point of view, and I mean really consider it.

> Yes, we have seen an improvement in the children's level of thinking. From my own classroom observations, the children are certainly more prepared to ask questions, and more prepared to discuss their thinking. In one class that I observed recently, it was quite lovely to see that one of the children was actually able to contradict something that the teacher had written, and she did it in such a nice way. It was quite a mature thing to do.

A link inspector said:

> The children are involved in their own learning, it's not just delivering it to them, but it's having them involved in the practical activities. And some of the practical activities have been so very carefully thought out, they are just superb, you just have to stand back in amazement at the way the children tackle them. It is having a different view

point on learning, viewing it from a different side, and it is this many faceted side of learning that is so good to see, together with the confidence that children have in the way that they answer questions. I do think that their response skills have improved.

The teachers were interviewed about the impact of CA on their pupils during the *Let's Think!* sessions. After further consultation with the teachers, the learning skills identified were grouped into a number of broad categories. They identified crucial changes in the children's cognitive skills, such as analysing, reflecting, thinking about thinking, problem solving, and formulating arguments. Changes in the children's communication skills (discussing, explaining, questioning, looking, listening, and extending their vocabulary) and social skills (turn-taking, team work, co-operating, and sharing) were also attributed to the CA programme. Not surprisingly, at the start of the year the teachers were more aware of the rapid development of the social and communication skills than they were of the cognitive changes, which were perceived as the slowest to develop over the year. Other highly valued changes which teachers attributed to *Let's Think!* were the way in which the children started to take responsibility for what they were doing, developed a sense of ownership of the group process and showed increasing independence. The teachers felt that the children were more actively involved and 'on-task' during *Let's Think!* activities than in other classroom work, partly because they enjoyed the experience so much. Teachers were often surprised at which children performed well in *Let's Think!* sessions. For example, less able children and normally quiet children, who were often those with poor literacy skills, did unexpectedly well and children who were still learning English were able to demonstrate ideas.

The teachers also said that the impact of the sessions on their children's learning skills extended to other areas of the curriculum, in particular mathematics and science. The skills that they particularly noted were their responses to questions, their increased awareness of what was being said by others and their greater co-operation and independence.

Teachers' Pedagogical Development

From the written evaluations, 100% of the teachers said that their teaching had changed as a result of *Let's Think!*, and that they used teaching strategies developed through this programme in other curriculum areas. In both these evaluations and in the interviews they particularly identified changes in their questioning techniques, as well as talking more generally about bridging content and methods across the curriculum. Many mentioned the fact that *Let's Think!* had made them concentrate on the processes as well as the results.

I think that each group is so different . When I listen to the children questioning, it is very challenging for me, as a teacher, to think about how I am going to answer their questions . It is challenging to get them to reflect, to get them up to a high order of thinking and then to get them to ask questions. You have got everything going on, you are trying to get them to work as a group, to think about questions and to think how they arrived at their answers.

> I think it is the fact that the children realise that, just because the teacher is sitting in the group, the teacher is not in control and not in charge. I start my sessions saying who is in charge and why and 'we all are'. This has followed through to the rest of my class teaching, the fact that I won't have all of the answers just because I am an adult. As a teacher you are so used to sitting and giving, I found it so hard to keep my mouth shut.

> Now, I'm not so worried if they don't get the answer, whereas I was before. This is because it is more about the process than the end result, isn't it?. When I started I thought that they had to get the answer, otherwise I wasn't doing it right.

Here are some comments from headteachers:

> *Let's Think!* is more focused and I think that the bonus of that is that it gives children the opportunity to work things out for themselves. I think that we don't do that sufficiently. One of the really good things that has come out of *Let's Think!* is to see that if we give them long enough they will sort it out. I have also seen the links that they make to solve problems and how they collaborate with each other. I suppose it sounds as if I am getting a bit carried away with it, but the *Let's Think!* children have become so tolerant, it's one of the nicest classes we have in the school because they are so tolerant of each other, so kind and they work so well together.

> In CA the value of the intensive group situation is so important. I think that what happened with the introduction of the National Curriculum is that this type of time with the children went out of the window and we suddenly accepted the 'whole-class' scenario. So I think that what *Let's Think!* is doing is saying that here we have half an hour, and when that time begins I will give my attention completely to you, with no interruptions whatever happens. It is actually giving them the time to talk, which is so important.

A link inspector said

> As I have seen more *Let's Think!* lessons I have … looked more carefully at how teachers' questioning skills have improved and certainly this has been the most noticeable outcome in the schools that I have been in - it's the teachers' abilities to get children to think for themselves, rather than giving them the answers.

In both written evaluations and in interviews, 100% of teachers linked the development of children's learning skills to changes in their own pedagogy, both in *Let's Think!* sessions and in other curriculum areas. Specifically, they identified changes in their questioning techniques, such as asking open-ended, probing, questions and asking children to explain their answers. They also noticed that they listened to their children more carefully and allowed more time for discussion. Overall, they increasingly saw themselves as facilitators, being less didactic and more able to sit back and give the children time to think and answer questions. Many also mentioned that *Let's Think!* had made them worry less about the end results and concentrate more on the processes.

The NQTs' experience

Evidence from interviews and from discussions which took place during centre-based PD days suggested that problems that all teachers had to resolve, and that most dealt with in the first half term, were still a concern for the majority of the 6 Newly Qualified Teachers (NQTs) in January. These included the timing of *Let's Think!* session in the day, organising the mixed ability groups, fitting half hour sessions into a crowded curriculum, and encouraging the rest of the class to work productively and independently. However, by June the majority were much more positive, although two of the three teachers who were still having difficulty were NQTs.

> It has taken quite a long time to get to grips with the children's behaviour and the management of the class when I'm with just a group. At the beginning I had to kind of push myself away from the rest of them, because it was new and I was new as well. Because it is half an hour, I was finding it very difficult to fit into the timetable, specially towards the end of term. Something happens in the afternoon, and I have got extra things to do, I find its very hard to fit it in. That is one of the main problems I've had with it really.

> I think time juggling has been difficult for me as an NQT, but I'm learning as I go along. I'm realising that if I do a *Let's Think!* activity maybe then I can do literacy with the other children, but it's really juggling. But then it's been really good on the training days because I've been able to talk to the experienced teachers and ask them what they do to get round the timetabling. I think at the beginning it was more difficult because I'd never had a class myself, but now this has given me almost more confidence.

> (*January*) It has been a lot and I have to be honest and say that for the first term I don't think that it was my priority. I had so much else to take on board, the kids, the class, the school, so *Let's Think!* did take a back seat, but I have done it and I thoroughly enjoy it. I enjoy the professional development days, I find them really good and then I come back and put it all into action and I am getting better.

> (*Same teacher in June*) I enjoy it so much more now, so in that sense it is working a lot better and they enjoy it a lot more. I don't know what has happened over the last term, we have all just suddenly gone "Oh this is how it works" and the whole class is working better, they are working independently now, they all know what to do. It took them and me a long time to get to grips with it, but management and things like that are a lot better, so it makes *Let's Think!* easier for me and they enjoy it so much more. I am much more flexible now, maybe that is what it is. I can relax, I can say I won't do that now I will do that tomorrow.

The headteachers were divided in their views about the participation of NQTs in the programme but all except one of the heads who had NQTs in their school were positive about the benefits. For example:

> I am very happy that we are lucky enough to be involved in *Let's Think!*. I think that the staff are benefiting from it and I can see that the children are benefiting even more.... and with an NQT in the class, I have been doubly impressed.

> The NQTs all felt they had benefited, but I felt they would have benefited more in their second year because in their first year they are really coming to terms with so many things in the classroom. They have to get to grips with so much more in the first year that I actually think they would be a lot more reflective on their practice in the following year.

The link inspector who covered schools with NQTs was also positive about their participation where the school provided good support.

> I was really impressed with the way the NQTs were doing *Let's Think!*, particularly where they have good systems in place already and they have another teacher who has already been involved in the project to support the NQT.

Evidence from the interviews with the teachers and from their discussions on the PD days suggested that the CA pedagogy of NQTs was slower to develop than that of the more experienced teachers. However, a detailed comparison of their questionnaire responses, where they were required to rate their CA pedagogical skills, found no significant difference between the NQTs and experienced teachers (independent t-tests, $p<.05$).

It is possible that the perceptions of the NQTs about the impact of CA on their children and on their own pedagogy changed over the year as their own standards about what was successful altered. For example, initially, they might have been very impressed with the contrast between the children's communicative skills in small group *Let's Think!* sessions and in whole class discussions. Later in the year, the demonstrations and feedback sessions with their teacher-tutors could have modified their expectations.

Overall, although there were mixed opinions about the inclusion of NQTs in the programme, the perceptions of those directly involved were that NQTs and their children do benefit from the programme (particularly where there is good support within the school and where the school is already involved in CA).

Reflection on Practice

In each year of the programme, teachers were asked to reflect on their *Let's Think!* sessions by writing a 'learning log'. Analyses of the 2000-2001 logs indicated that only 2 teachers found that it was useful, while 9 teachers said that it was not useful. These responses were in the same order as those given by teachers in the previous year (23% and 69% respectively). In contrast, 71% of teachers in the first year of *Let's Think!* (1998-99) found the log useful and none said that it was not useful.

Teachers in the first two years listed many ways in which they found the log useful, both as a reminder of the activities and as a record of how their children had developed. The log also served to keep the teachers focused on their aims and on changes in their pedagogy. Even where teachers in the first two years said that they did not have the time or the motivation to fill it in, they commented that they could see that it could be valuable. The 2000-01 teachers made many fewer positive or negative comments about the log, perhaps because there was less emphasis on its use during the PD sessions. The few comments that were made included a preference for discussions, the need for more guidance on what to record, and a format that would make recording more manageable.

An important aspect of *Let's Think!* is that the teachers as well as the children should reflect on the activities, but they need to be able to do so rapidly because the pressures on their time are so great. One possibility suggested by a number of teachers (36% in 2000-01) was that some sort of formative evaluation sheet to be filled in at the end of each session or at the end of each week would be useful (it could also help the teachers identify areas to work on with their teacher-tutors).

THE SYSTEMIC PROFESSIONAL DEVELOPMENT PROGRAMME: THE TEACHERS' EXPERIENCE OF WORKING WITH TEACHER-TUTORS

For the first time (2000-20010), teachers on the programme were partly tutored by experienced *Let's Think!* teachers acting as teacher-tutors. In January, some teachers were unsure about the role of their teacher-tutor in relation to the role of the borough co-ordinator. When they were asked about the work in schools, over half mentioned spontaneously how much they valued the input of the borough co-ordinator and how they modelled their questioning on hers. By June, all of the teachers were much more aware of the role of the teacher-tutor and only three teachers (18%) mentioned the contribution of the borough co-ordinator.

Although 90% of the teachers had no choice about joining the programme and 60% had never even heard of Cognitive Acceleration at the start of the year, the overall professional development programme (centre-based and school-based work) was described by 100% of teachers as valuable and highly motivating. When the teachers were asked what advice they would give to the next year's *Let's Think!* teachers, a typical comment was "make the most of this professional development, opportunities like this don't come very often"!

Centre-Based Professional Development Days

The teacher-tutors' contributions to the centre-based professional development days included simulating activities (demonstrating without children), leading discussions and delivering occasional individual sessions on *Let's Think!* pedagogy and organisation. The days were described as "vital" and "inspirational" by many teachers in their interviews. Examination of the written evaluations from the last three years showed that the 2000-2001 teachers were as positive as teachers had been in previous years about *Let's Think!* in general and about the centre-based days in particular. In each of the last three years 100% of teachers who had attended these days felt that they had provided them with valuable professional development.

Centre-based simulations of activities and school-based demonstrations were cited by 86% of teachers as some of the most useful aspects of the PD programme. They were clear that it would not have been sufficient just to read about the activities and that it was crucial to see them demonstrated or simulated. Modelling the teacher's language, particularly questioning techniques, played an important part in increasing teachers' confidence to carry out activities. In 1999-00, all simulations were given by the borough co-ordinator, whereas in 2000-01 many simulations were carried out by the teacher-tutors. After these sessions, as part of the evaluation, teachers rated how confident they felt about carrying out the activities, using a 0-9 scale (0 = least to 9 = most confident). Median scores were calculated for both years and comparisons of mean scores were made using independent t-tests where ratings were available for both years. Results are shown in Figure 9.6 for tutor demonstrators and 9.7 for demonstrations by the borough co-ordinator.

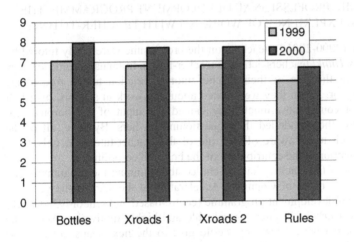

Figure 9.6 Teachers' confidence in using activities: tutor demonstrated

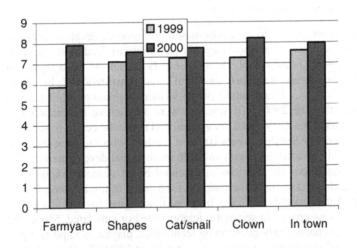

Figure 9.7 Teachers' confidence in using activities: borough co-ordinator demonstrated

Activities demonstrated on the last two professional development days were rated more highly by 2000-01teachers than by the teachers from the previous year. The differences were significant for the Bottles, the Farmyard and the Clown activities ($t=2.98$, $p=0.006$; $t-5.33$, $p=0.0001$; $t=2.17$, $p=0.039$ respectively) and they approached significance for both of the Crossroads activities ($t=1.89$, $p=0.072$ for both). The 2000-01 teachers particularly valued the demonstrations by teacher-tutors and rated themselves equally confident to carry out activities demonstrated by either teacher-tutors or by the borough co-ordinator (no significant differences were found, $p>0.05$). The greater confidence shown by the 2000-01 cohort of teachers was not surprising because the activities were still being modified and refined during the previous professional development year. In September 2000, this group were less confident than the previous year's group had been at the beginning of their year.

Group reflections were also a vital component of the centre-based PD days. These were cited by 76% of teachers as a very important aspect of their professional development. The opportunity to discuss *Let's Think!* with others who were experiencing similar problems and with more experienced CA teachers (the teacher-tutors) provided essential support.

> What is really useful is getting together with everybody else and being able to talk about the problems. It's very much something that only a small group of people know about, so I can't necessarily talk about it to anybody here (*in her school*). They are all very supportive, but they don't understand it in the way that other *Let's Think!* people understand it. That support, it really lifts the weight from your shoulders.

> The groups, the discussions, the fact that everybody shares things. People aren't backwards in coming forwards and everybody has got a part to play. It really good that people have suggestions to overcome your problems

> (*The PD days*) are really, really good. There are a lot of very competent people there and everybody is very into *Let's Think!* If you've got a problem there will be somebody there who has found a solution to it and I find that very, very helpful. There are an awful lot of very knowledgeable people, a lot of people who have done CA and had similar problems, and that I find very helpful.

Theoretical input on the centre-based days was cited in the written evaluations as one of the most valuable aspects of their professional development by almost half the teachers (43% in 2000-01 and 53% in 1999-00).

> Obviously I couldn't have done it without the PD days. All of it is valuable really, but particularly the theory and actually how you do the activities. Without the theory it wouldn't really mean that much, you would just be doing it by rote.

> I couldn't have done it without it (*the PD*) because I wouldn't have understood the five pillars, and I do understand them now. But it is the usual thing, theory and practice are two different things and it wasn't until I got in with the kids that I thought "how am I going to tackle this?". It's because we've been over it and discussed it, that I can now say "that is what they mean by concrete preparation".

Twilight Sessions

Two optional twilight sessions were organised by the teacher-tutors in the summer term, held in schools rather than in the borough's PD centre. 60% of teachers attended one of these sessions. In spite of this, over half of the teachers reported that the lack of centre-based PD days had left them feeling isolated in the summer term. All of those who attended a twilight session valued the chance to discuss *Let's Think!* with both teacher-tutors and other teachers and were very positive about the continued support and encouragement that they received at the meetings. Both meetings had an informal structure, with agenda that were determined by all of the participants. An offer by the teacher-tutors to demonstrate activities was not taken up but, in retrospect, a third of the teachers felt that these sessions should have been more structured and that they would have benefited from demonstrations.

> I think that because of the twilight session I still feel very much part of the *Let's Think!* group. If I hadn't had anything then I would just have been doing the activities and that would have been it. I really enjoyed having that contact.

> I must admit that, because the PD days had stopped, I was feeling a bit isolated in the classroom. I was even wondering if the others were still doing *Let's Think!*, so it was actually quite nice to go to the twilight. I do think that to meet up with everyone in the summer is important, because it is a very long term. It was great to go there and chat, but if we had had some sort of agenda to follow it would have been a more focused and productive meeting.

> It was good because it was casual and we could discuss how we were going on, that sort of thing. It was useful to have something at that time, although it wasn't as good as a full day. However, if you are going to have these in the future, instead of some of the PD days, then maybe they will have to be more structured.

School-based Professional Development

Overall, this work was described by the teachers as a well organised and a helpful aspect of their professional development, although 22% of the teachers mentioned some aspects that could be modified or improved. Mostly they wanted more PD sessions and meetings with their tutors.

The teacher-tutors demonstrated selective *Let's Think!* activities in the teachers' classrooms, mainly in the first term. Teachers reported that it was very useful to see an experienced CA teacher demonstrating the activities with their own children because it allowed them to watch individual children and to observe how they and the teacher-tutor reacted. Teachers were likely to give teacher-tutors their most difficult groups for demonstrations.

> Demonstrations are more helpful than anything, because they are actually specific to your class and because whoever it is who is demonstrating can choose your most difficult group and you can see what they do.

It was helpful to see someone else doing it, and to realise that you are not expected to be perfect. There was a little girl who didn't say much when she was doing it with the tutor, so it wasn't just me. OK, fair enough, some children just do not say that much.

Yes, I sit back and watch them and think, "hey, you don't answer like that when I'm doing it", because they are different, they respond differently. Its is good, really good to have them come in, especially to demonstrate.

The teacher-tutors went on to observe the teachers' own *Let's Think!* sessions and to give feedback to the teacher concerned. The majority of teachers seemed to be happy to be observed, although one NQT said that being observed was 'scary at first' and another teacher mentioned that it was embarrassing because the *Let's Think!* session observed by the teacher-tutor was 'really bad'. Teachers thought that the early observation and feedback sessions were valuable because they helped to boost confidence and reassure them that they were 'on the right lines' and that they were making progress.

Well yes, it was helpful sometimes not just to have criticism, but to be reassured. I was a bit worried that my groups weren't taking it on, the quality of talk didn't seem to be that great to me. I think it really took till the end of the first term for any kind of exploratory talk among the children to emerge at all. So after a couple of the observations it was nice to talk to someone and be reassured that it was a process we had to go through, and to be able to recognise the progress that they had made. So it wasn't just the criticism, it was a bit of reassurance as well.

Most teachers were very positive about the coaching done in the later observation and feedback visits. They clearly found the feedback relevant and justified and they also valued sharing ideas with their teacher-tutors. The teachers reported that the specific issues raised by their teacher-tutors' feedback included altering their questioning style, stretching the children to a greater extent and stimulating metacognition through questioning.

I think these sessions have been very successful in terms of having someone, who has done it before, saying "maybe you could try this, maybe you could try that". It's on a more on a personal basis than when it's done on a PD day, which is very much to do with what can help all of us.

I think the observations have been useful in terms of my confidence. I wasn't totally sure that I was getting it right, but then I was told that "yes, they are working well as a group". And the feedback was good, in that I knew that I hadn't been asking metacognitive questions throughout that session, and that was picked up. And it wasn't an awful thing to be told, because I was aware that it was a fair criticism...

I think when I was first doing *Let's Think!*, I was doing the activities without actually thinking about it in theory terms. I wasn't thinking about it in chunks under the headings such as 'social construction'. I have found it much more helpful now that I can put the two together. I can say to myself "right, I need to think about asking them if this reminds them of anything" so that they are doing some sort of bridging. I need to ask them at the end what they found difficult, or in the middle of it, "why are you finding that difficult?". Just to actually be able to think about it more in terms of the theory, rather than just doing the activity, that's why the observations have been helpful.

Some teachers recognised the feedback strategy of identifying good points first and then picking up on just one or two points for development.

> When my teacher-tutor came in it was good, lots of positive feedback and just a couple of things to work on. That was helpful because sometimes you don't realise what you are saying or what you are actually doing, unless someone says "that was a really good question".

> ....but she said .... that the metacognition wasn't strong enough, but she praised the good bits. You know what I mean, the 'shit sandwich' as they say.

The main area for attention that was highlighted by the January interviews concerned the teachers' perceptions that some teacher-tutors may have been too diffident about identifying areas that needed attention and giving constructive feedback. However, other teachers did mention how much they valued the supportive and constructive feedback that they had received. Initially, there was an emphasis by most teacher-tutors on oral rather written feedback, and there was some indication from the interviews that this was what the teachers wanted. By the second interview more teachers had received written as well as oral feedback but 5 received only written feedback. The need for more discussion time after an observation session was commented on by a number of teachers. This problem stems largely from the fact that in most primary schools it is difficult for teachers to take time out of their classroom work to talk to their teacher-tutors. This was a problem which was also identified by teacher-tutors who had limited flexibility when timing their visits to schools.

> I can't think of how the observation and feedback sessions could have been made more useful, not really, no. Apart from having more time to discuss it afterwards, but that's to do with release time. You could try and have ten minutes release or ten minutes at playtime by arranging it so that the tutor is in just before playtime.

> I think that with the discussion, the difficulty is really just the logistics. I am straight back into class because I haven't got anybody to take over. I think that having time to discuss it is always going to be a problem unless you go straight into a break. And I haven't really spent time with her when I have gone to meetings, we have gone straight into groups. I suppose that I could have seen her at lunch time, but we have never done that.

The teachers were encouraged to make 'peer observation' visits to *Let's Think!* colleagues in other schools. Headteachers were given money for two days supply cover in order to allow each teacher to observe the *Let's Think!* sessions of up to four other Year 1 teachers. Sixty percent of the teachers made at least one visit. However, seven teachers made no visits and seven received no visits, with six teachers neither making nor receiving a visit. Teachers who organised themselves into small groups of three or four and then arranged reciprocal peer observations were the teachers who made and received the most visits. Reasons given for not making visits or making fewer visits than desired were numerous. Some teachers (or their headteachers) felt that they had already had too many days out attending other courses. In other cases, absence through illness meant that teachers either missed the planned visit or did not wish to take more time out after they had returned. Some found that it was difficult to arrange the visits with other teachers or that a booked supply teacher failed to turn up.

> I just found myself really flat out and couldn't take the time out of class, it was just like 'I can't'. ....

> It was very useful but hard to co-ordinate, though I might have been a bit lazy. I found it very hard to even say "can I come and watch you?".

> There were a couple of problems in organising the peer visits. A couple of schools wouldn't let us. One of them said "one of our Year 1 teachers is away and the other is too busy, so she can't do it". For next years group I would team up people, not just make them do it themselves.

All who participated found them a very positive experience. Two general benefits were mentioned. Teachers were able to see other Year 1 classes, to see how they were organised and to share ideas on this. They were able to compare other classes with their own and gauge how they were getting on with their own classes. Many stated that the visits also provided support and encouragement and helped to boost confidence.

A number of specific benefits relating to *Let's Think!* teaching were also reported. Teachers were able to see a variety of ways of using an activity and how different children responded in the same activity. They reported that it was valuable to see more experienced teachers teaching *Let's Think!*, and particularly mentioned modelling of questioning styles. They valued discussing different ways of encouraging all children to participate with other group members and many noted that they were able to pick up ideas on what to do with the rest of the class while running the *Let's Think!* sessions (e.g. creative writing in *Let's Think!* diaries).

> Oh, it was good because I did three in a row, and got to see this one activity that I hadn't done yet, and the three people did it so differently, I was amazed. It was really interesting to see how differently they all did it, and to see how the children responded, even though they are different children doing it...

> ....just to see the way other teachers teach, to see if there is anything I could do better or differently. I was also interested to see how the children are, because I think my class were very, very good with *Let's Think!* and I wanted to see whether it was me, thinking "oh they're fab", or whether they really were good at the activities. So I was quite intrigued in that sense.

THE SYSTEMIC PROFESSIONAL DEVELOPMENT PROGRAMME: THE TEACHER-TUTORS' EXPERIENCE

The headteachers, link inspectors and the majority of teachers felt that the teacher-tutor scheme was a good idea that had worked well in practice. A typical view of the way that experienced *Let's Think!* teachers were used as teacher-tutors was given by one of the link inspectors:

> The teacher tutor scheme is a good model of dissemination, and teachers have confidence in other teachers who they know have credibility. The feedback that I have had is all positive.

Teacher-tutors' perception of their own role

At the start of the year, most of the teachers expected the teacher-tutors to adopt the traditional tutor role of starting with expert demonstrations and moving on to observations with feedback. However, the teacher-tutors did not necessarily see their role in these terms, with most of them having reservations about being seen as 'experts' in any sense. There was little difference in age and general experience between the teachers and the teacher-tutors, so that they saw each other as peers. This, rather than any ideological notions of coaching, may have generated the teacher-tutors' uncertainty and their reluctance to be seen as experts.

> In a way, at the start, I almost avoided my tutees because I felt awkward, I didn't know where I stood. When you see the other tutors, you know where you stand, but when you go in to other teachers' classes, you feel that you are supposed to know more than everybody else.

> To start with I think that they still preferred to hear things from (the borough co-ordinator), the authority thing, but they liked our demonstrations, and as we went on it got better.

> Beforehand, I was thinking that there was no reason why they should listen to me, because I was just one of them, as it were. And although I am not saying that we had 'authority' in any great sense, I was quite surprised by the fact that there was a sense that we had got more experience than they had, so they listened to us. I didn't feel that they would only listen to what (the borough co-ordinator) said, because she, and nobody else, was the fountain of all knowledge.

As the year progressed, the fact that the teachers regarded them as experts began to influence the teacher-tutors' view of themselves. By the end of the year, the teacher-tutors had all decided that their role was both supportive and instructional. They saw themselves as practitioner tutors and all felt comfortable in giving what could be seen as a traditional form of feedback. They all felt that they had made a difference to the teachers' professional development.

> The tutor's role is to do with supporting the teachers and developing their expertise, and that is a difficult skill. It's not just being the teacher's friend. I am sure that in years to come, I will still be developing the skill of being able to pick out that one thing that you know they can develop, and that you can help them to develop, by reviewing their work and setting them a new target.

> The main role is constructive criticism. I'd also say keeping the positive atmosphere of *Let's Think!* going, encouragement, essential qualities like that. We don't know everything, but encouraging teachers to reflect on how they are doing it, for example by referring to the challenge then questioning, can be very helpful.

The Impact of being a Teacher-Tutor: their own Professional Development

The teacher-tutors were clear that their own tutor professional development sessions together with the experience of tutoring had greatly improved their own personal and professional skills. All talked about the improvement in their professional skills outside the classroom and most felt that their pedagogical skills in their classroom had benefited as well

> It does all have a positive impact on my own professional development and sometimes I come away having been reminded of something that I have got to improve on in my own classroom.

> I really enjoy being a tutor and talking at the CA convention about the impact of *Let's Think!* on my teaching style has made me more aware of my teaching. Now I reflect on how I teach a lot more.

> There have been numerous initiatives that have come in whilst I have been a teacher but it is definitely *Let's Think!* tutoring that has had the biggest impact on my teaching career. Certainly much more impact than anything else, not just in terms opening up opportunities, but helping me to see my role as a teacher in a very different light to previously.

The Impact of being a Teacher-Tutor on the Tutor's Own School

Teacher-tutors had to take 10 days out of their own school to visit their tutees. Whether or not they thought that this had a detrimental effect on the their own classes was partly related to the arrangement that had been made for supply cover. There were fewer problems in schools where the same teacher was always used to cover the teacher-tutor's class and this arrangement was even seen as an advantage by some. However, where the supply teacher was different for every lesson, some teacher-tutors felt guilty about leaving their children or felt that it increased the pressure on them when they returned to their own school.

> I think that going out of my classroom has had an impact on *Let's Think!* because it is hard to keep everything going and to catch up.

> I have a very well behaved class but I don't like going out so often, it isn't very good, especially with SATs coming up.

> Well I'm very fortunate this year to have a class who are a lovely class. They are very amenable and adaptable and well used now to having different faces taking them.

Some headteachers felt that the teacher-tutor's absence would not be well regarded by parents, indeed when some teacher-tutors were scheduled to have a day out of their own school, their heads asked them to be in their own classrooms at the start of the day and to be back before the parents turned up in the afternoon. However, all headteachers did agree that the teacher-tutors, and therefore in the end the school, had benefited from the experience.

> The tutor is really enjoying what she is doing, and I think anything that really provides a very good teacher with the experience that *Let's Think!* provides is just amazing, particularly for young teachers. We have benefited from having a tutor in the school because, having been trained to observe staff and give them feedback has helped her in her role as curriculum co-ordinator. I think that giving feedback to colleagues who may be older or more experienced is not very easy, so I think that from that point of view, it has been very helpful.

> The tutor in our school has been very positive about the tutor training. She has been caught as a professional at the right time for her and I think she feels very empowered as a professional. I think it is great for young teachers to have this opportunity.

> The problems have been mainly to do with the cover arrangements for the work in schools. We can usually manage the cover for the PD days, but the parents feel very strongly that they want their children to be taught by a teacher that they know. They want to see that teacher there at 9.00 in the morning and again at going home time, and I think that she has been a bit upset by it. However, this has been very powerful staff development for her which obviously has had a cognitive effect on the children and the staff within the school.

Predictably, teacher-tutors felt that whether being a *Let's Think!* tutor had any impact on the whole school or other staff in their school very much depended on the attitude of their headteacher. Although the interviews with the heads showed almost universal support and enthusiasm for the teacher-tutor scheme, some of the teacher-tutors felt that in reality their head's support had been less than enthusiastic. This may have been partly due to the fact that, while the *Let's Think!* programme was a major part of the teacher-tutors' professional experience, it was a relatively small part of the headteachers' experience of running the whole school.

> If you haven't got the head behind it you are fighting a losing battle. It's lovely when I have those PD days and talk to the other teachers and tutors, but when I come back into here, it is a big school, and the emphasis tends to be on Key Stage 2 so it can be quite difficult. But because I feel so keen about it, and I know there are other new, young teachers who are very interested I arranged a staff meeting myself. I really enjoy being a tutor, but when you come back to the whole picture of your school life it can be frustrating.

> Nobody, not even the head, has ever been to observe what I am doing in school, I did run an INSET session but there isn't a lot of interest I can tell you. (The borough co-ordinator) has come in and done days, we have really tried, but I think that it has to come from the top and it is hasn't.

In addition several teacher-tutors were ambivalent about the attitude of colleagues in their own schools

> I don't know what my colleagues think about me being a tutor. They joke that I'm never here or that I'm a part-timer. It never appears to be in critical way and I haven't had anything critical said to my face, but I think that there is this feeling... The headteacher

is very supportive, I think she views it that I'm receiving training that is useful not just for *Let's Think!* but in other areas of the school life.

Overall more than half of the schools had organised meetings where they demonstrated and talked about *Let's Think!* to their colleagues and/or to the governors of the school. Predictably, these tended to be in the schools where the headteachers were fully supportive of CA.

Teacher-Tutors' Views of the Professional Development Programme

Teacher-tutors expressed views on all aspects of the teachers' professional development programme. Two important issues that they raised, which related to the centre-based days, were maintenance of a strong group identity among the *Let's Think!* teachers (something that teachers had also noted as an important feature of this programme) and the amount of theory that should be included (particularly as they knew that in the following year there was to be a reduction in the time allowed for the professional development).

The professional development days have always included time for the teachers to talk to each other. Most of this talk is focused on particular *Let's Think!* issues but at other times the sessions are more loosely organised to encourage the teachers to get to know each other and to identify and develop a common set of ideals. This deliberate attempt to encourage and establish a group identity between the teachers and between the teachers and tutors was identified by tutors as one of the successes of these days.

> Maybe on some occasions when the teachers are chatting together it's not focused, but in my opinion, the fact that teachers are bonding together, creates a greater sense of ownership over the programme. You feel that you get on well, and that you are working as a team, and it is so important to have that cohesive group of people. I think, the fact that everybody who has gone to these PD days is really very positive about *Let's Think!*, has a lot to do with the fact that they, as a group of people, identify with each other and know something about each other, and know about each other's schools and timetables. I think that is where the difference is, between these PD days and other courses you go on. You do get great content from other courses, but because they are only one day courses, you don't have that sense that you are all there as a group.

> Talking to the teachers, they value that opportunity to talk, to feed back about how it is going. It's great that has been held on to and that time is allowed for people to voice their opinions about it all. That is very good for people's motivation.

In one sense, the expertise of the teacher-tutors lay in the details of the practical delivery of *Let's Think!* in the classroom. However, their own CA professional development had emphasised that an understanding of the theoretical underpinning is essential to successful implementation, and that theory and practice cannot be separated. This was reflected in the views of the teacher-tutors who all felt that it was essential for the professional development days to include the theoretical aspects of CA as well as the practical.

.....it is really important that there has been some theory input........

> When we were trained we had lots of different types of theory sessions, lots of different kinds of input throughout the year. Some people take it on board straight away and will always go off and do their own research into it, but there are some who would never read anything outside these sessions. It is essential for them to keep getting it.

Most teacher-tutors mentioned the theory input on their own tutor professional development days. They saw it as essential that this was included so that they could refer to it on both their school visits and the teachers' professional development days.

> The bit that I have always enjoyed was the theory and that was something that was a great feature over the last two years. We have had in-depth theory on our tutor training days and I have enjoyed that. It may not necessarily have been directly related to our role as tutor but it is all useful background that I really value.

> .....even if there has to be something cut out of the tutors' PD next year you have to hold on to giving the theory.....

> Maybe these days could be improved by spending even more time talking about theory, so that if you were doing it in a PD day, you are keeping abreast of it.

DISCUSSION

The overriding impression gained from this research is that this professional development programme involving teacher-tutors was perceived as exceptionally successful in enhancing teachers' pedagogy and children's learning skills. The Link Inspectors and headteachers reported that the teachers were enthusiastic about *Let's Think!* and competent in teaching it. The teachers also confirmed their enthusiasm for it, with more than 70% of them rating most aspects as successful. The enthusiasm and competence of the teachers could not be accounted for on the basis that they were a self-selected group, who had chosen the course. None of them had any choice about participating in the programme and almost two thirds of them had never heard of Cognitive Acceleration before they started. Schools were selected by the borough staff because the funding for this programme came from the Single Regeneration Budget. So why was it so well regarded by all those involved? It's success must be attributed to the specific aims and methods of *Let's Think!*, the design and implementation of the professional development programme, and the enthusiasm, skills and commitment of all those who were involved.

The aims and methods of the *Let's Think!* programme were immediately seen by teachers as being compatible with, and complementary to, their own teaching objectives. Although they were particularly enthusiastic about the specific cognitive goals, they also valued the broader social, communicative and motivational benefits of the programme. The chance to work with small groups allowed them to observe closely the development of individual children and this was particularly welcomed as an important complement to their normal classroom teaching. They also valued the fact that, in the *Let's Think!* sessions, the process of resolving the challenge was more important than the final result. Many were delighted that the absence of any requirement to produce written evidence of their work gave some children, particularly those with poorer literacy skills, an unexpected chance to demonstrate their ability and to boost their confidence. In short, the aims, methods and outcomes of *Let's Think!* were congruent with the teachers' beliefs, motivations and expectations.

The design and implementation of the professional development programme included all of the components that have been identified by researchers as important for the success of this sort of curriculum innovation (see chapters 10 and 11). For example, it was founded on a clear theoretical framework and included demonstrations in simulated and classroom settings. It also required the teachers to practise extensively the innovation in their classrooms, and included in-class observation with feedback and support from both peers and experts. The inclusion of teacher-tutors in the professional development programme also contributed to the success of the scheme. The new *Let's Think!* teachers valued the contact with experienced *Let's Think!* teachers, some of whom had been directly involved in the development of the activities and the classroom methods. This contact emphasised the collaborative and teacher-focused elements of the professional development. A more general factor, which could also have contributed to the positive responses of the teachers, is that Year 1 teachers have relatively few opportunities to meet and discuss their pedagogy with first year teachers from other schools. The centre-based professional development days were therefore important, not only for discussing *Let's Think!*, but also for sharing ideas and information and providing mutual support. This was reinforced during their peer observation visits to other teachers' schools.

A number of other factors which affected the success of the professional development programme were identified. As might have been expected, NQTs took longer to develop their CA expertise than more experienced teachers and had more problems in the first term with classroom organisation. For these reasons, as a group, they needed more help from their teacher-tutors in the first two terms, together with continued support in the summer term. Most teachers did not make effective use of the learning logs and generally needed more guidance on how and when to reflect on their *Let's Think!* teaching. A more explicit 'reflection guide', which is easy and quick to complete, but which focuses the process of reflection, would have helped the teachers in this respect. The peer visits were particularly useful but not all teachers participated. There were a variety of reasons for their non-participation, some of which related to the teachers' own attitudes and organisational skills, and some to wider school issues. Finally, although almost all headteachers voiced their approval of *Let's Think!*, there tended to be more problems in schools where the headteacher did not actively support the programme.

The involvement of teacher-tutors introduced an element of peer coaching into the programme and clearly made a significant contribution to its overall success. Teachers reacted very favourably both to their centre-based and in-school contributions. In addition to valuing the chance to reflect on and discuss *Let's Think!* issues with experienced teachers, they found it particularly useful to be able to watch the more experienced teacher-tutors demonstrating activities with their (the teacher's) own children. The teachers were equally confident about carrying out activities that had been simulated or demonstrated by either the teacher-tutors or the borough co-ordinator.

The research literature (for example Joyce & Showers, 1995; Showers & Joyce, 1996; Yarrow & Millwater, 1997) had suggested that a range of problems could be generated when teacher-tutors gave feedback to other teachers. However, although some of these problems did occur, not only were they mainly resolved by the end of the year, but there were some particular benefits for the teachers that were only associated with the giving of feedback. There was no hint from the interviews or written evaluations that teachers felt that their privacy was in any way 'invaded' by a teacher-tutor demonstrating *Let's Think!* with their children, or observing one of their own sessions. Initially, some of the teacher-tutors did have worries about the possible hierarchical aspects of the job, which they thought cast them as experts who were observing, making judgements and giving feedback to teacher colleagues. Later in the year, the role had developed into one that most of them were comfortable with. The teachers' only adverse comment was that they needed more time to discuss both the demonstration and observation sessions with their teacher-tutors. Unfortunately, the logistics of organising the teacher-tutor visits together with the pattern of the primary school day made this difficult in many cases. The most effective solution, if the visit could not be arranged just before a school break, was for the teacher to organise someone to cover the class for a short time after the observation session. However this was not possible in many schools.

This school-based work of the teacher-tutors was relatively expensive to fund and obtaining reliable supply cover for their visits was sometimes a problem. The most effective solution would seem to be to develop further the idea of having a regular, known, supply teacher - preferably one who was knowledgeable about *Let's Think!*, who could take over the teacher-tutor's class. This may go some way to solve the fact that some teachers felt guilty about leaving their own classes with a supply teacher, although it would not entirely counter the view of one headteacher who was worried that parents would not approve of the teacher-tutor's absence from their children's class. The centre-based work of the teacher-tutors was rather more limited than their work in school, as they were only funded to attend two of the days. This meant that teachers were not always able to discuss *Let's Think!* issues with their own tutors on those days. Even where their attendance had been prearranged, in-school factors meant that they were not always able to be present. This created some difficulties when it came to integrating the contribution of the teacher-tutors. However, as a group, the teacher-tutors became more involved with the centre-based professional development sessions as the year progressed, eventually taking sole responsibility for organising and running two twilight sessions in the summer term. Following on from the example set by their own *Let's Think!* professional development, they placed great emphasis on ensuring that their beliefs and values and those of the teachers were congruent. Contrary to some research findings about other teacher-led professional development programmes (Wu, 1987) (and possibly because of the strong theoretical base to CA), they all emphasised the importance of including theory in these sessions. However, in this developmental year, much of the theoretical input came from the borough co-ordinator or university staff. Although this may be reduced in future years, some continuing contribution from them was seen as essential by most of the teacher-tutors, if only to keep themselves up-to-date and to refresh their own ideas.

A number of general observations about the professional development programme can be made. Firstly, what does seem clear is that the isolation of the Year 1 teachers and the teacher-tutors in their individual schools means that the complex organisation of the PD programme, including the teacher-tutor visits, can be carried out more effectively by a central co-ordinator, rather than by the individuals concerned. Secondly, the programme reported here was a reduced version of the 1999-2000 (developmental) one and subsequent programmes are likely to be reduced even further, in an attempt to reduce costs by finding a 'minimum architecture' for the professional development. However, all of the teachers and teacher-tutors urged caution when it came to reducing the programme. Some commented that although the twilight sessions were useful, they were not a substitute for centre-based days and others would like to have seen the centre-based and school-based PD continuing over all three terms. Certainly, there was evidence to suggest that, in the summer term, some teachers' motivation and their ability to reflect on their teaching would have been improved if there had been more contact with other teachers and with their teacher-tutors. However, while extending the professional development was not a practical option for financial reasons, the situation in the summer term could have been improved if, at the start of the year, the schools had been more definitely arranged into cluster groups for at least some or their work and a communications network had been set up.

Overall, as would be expected, the skills of the teacher-tutors took time to develop and not all problems inherent in such a programme were resolved by the end of the year. The teacher-tutors went on to be involved in the 2001-2002 professional development programme for the next generation of teacher-tutors (recruited from the 2000-2001 intake of teachers). An additional benefit for the borough was the retention of these experienced teachers, five out of six of whom said that they would have left Hammersmith and Fulham if they had not been involved in the scheme.

The underlying aim of the scheme was to offer a cost-effective method of continuing professional development within a borough. It seems clear that it was effective, but there is some question as to whether the use of teacher-tutors is necessarily very much cheaper – given supply cover costs and the continuing need for co-ordination and some expert input – than a more traditional system of buying in a complete PD team. For the time being at least, the borough decision is that the advantages of the scheme are well worth the cost, and the experience has provided a firm foundation for the establishment of a continuing systemic professional development scheme in Hammersmith and Fulham.

PART 3: MODELLING PROFESSIONAL DEVELOPMENT

10. RESEARCHING PROFESSIONAL DEVELOPMENT: JUST HOW COMPLEX IS IT?

In this chapter we will re-visit the question of what methods are available for evaluating the effects of professional development on teachers and on their students. This will entail a consideration of the epistemology of professional development, that is the nature of knowledge about professional development and how that knowledge might be constructed. We will consider some arguments from a postmodernist perspective about the dangers of over-determinism in methods of evaluating professional development and in particular the viability of 'process-product' research. Finally we will critically consider the various methods that have been described in part 2 of this book and draw some conclusions about effective (including cost-effective) methods of assessing the effects of particular PD programmes.

Before embarking on this enterprise, it may be worth recalling from chapter 1 the relative sharpness of focus of the enterprise of this book. We are concerned here not with the evaluation of school effectiveness or school improvement in general, let alone the evaluation of national educational change, but more modestly the evaluation of professional development of teachers. To be sure, PD cannot be isolated from its school context, but as we bring our lens to bear on the PD we will treat the school (and local authority) context largely as a given environment. Changing this wider environment may be necessary, but would be the subject of a different book.

THE EPISTEMOLOGICAL NATURE OF PROFESSIONAL DEVELOPMENT

The nature of professional development for teachers relates directly to the nature of teaching. Atkinson & Claxton (2000) have usefully characterised notions of teacher education each of which has been fashionable and has had its adherents over the past 40 years. These range in their dependence on conscious theory-driven action, from simple apprenticeship ('learning from Nellie'), through the unguided reflective practitioner and craft-skill ideas, to scholastic rationality based on the foundation disciplines of philosophy, psychology, and sociology. This spectrum has an echo in the age-old debate about teaching as an art or as a science (Gage, 1978). At one extreme, the story seems to be that the process of teaching cannot be taught in any direct way, it is simply too subtle and complex, and too dependent on deep-rooted talent and personality to be amenable to significant improvement through instruction or practice. At the other, technical-rational, end teaching is a skill based on sound theory and well-established general methods which simply require diligent study and plenty of practice in their application to particular situations. Stoll & Fink (1996) suggest that the idea of the teacher as a quasi-professional or as a skilled tradeseperson is associated with a transmission-sequential notion of knowledge, such that the teacher takes what is given to him/her by curriculum boards and textbook writers, and 'delivers' it efficiently.

But where along the route from one of these untenable extremes to the other do we pitch our camp? In Atkinson and Claxton's book referred to above, Furlong (2000) argues that the postmodernist relativistic interpretation of the notion of learning as a reflective practitioner falls into a trap similar to that of the old apprenticeship model, by under-representing what is known about teaching and placing too much reliance on the subjective truths of the individual without reference to external validation. On the other hand the scholastic approach and demands for teaching methods to be evidence-based (e.g. David Hargreaves, 1997) place too much faith on the reliability of objective truth. Furlong argues that teacher education should steer a middle way between these two 'flawed truths' and rely rather on critical discourse which continually questions underlying value judgements. The process of open and informed questioning itself drives the development of well-founded professionalism. It is not that there are no truths, nor that there are established truths which can be used off the shelf, but that any truths that there are need to be personally reconstructed and continually tested against reality.

We can now explore how such a middle way approach might apply to the nature and methods of professional development. Consider this series: astrology - meteorology – economics – professional development – Newtonian physics. Each is an example of a field which has certain characteristic methods of study and expectations of possible determination and control. We can dismiss the first quite quickly. Astrology uses a pseudo-scientific language such as 'alignment of planets and stars' or the entering of exact date and time of birth (why not of conception?) together with carefully vague statements of outcome to gull the simple-minded into the belief that there is some connection between the relative position of stars and human behaviour. There is in fact no connection at all between the input variables of astrology and any measurable outcome. (At the same level of determinism I am reminded of Eysenck's comment on psychoanalysis:

> By invoking the mechanisms of reaction and compensation, any outcome can be attributed to any cause)[1]

---

[1] This quotation is burned in my memory from my days as a young researcher, but now I have failed to locate a proper reference for it. The sentiment is fully spelled out in Eysenck (1953) pp235-235

The next two in the series, meteorology and economics, each have enormous and obvious importance for the physical safety, social stability, and financial well-being of society and consequently considerable effort and funding has been devoted to trying to understand them. In both cases increasingly sophisticated mathematical models are built to try to establish some predictability of outcomes, with more and more input variables being postulated, weighted, and entered into the equations. Everyday experience of the weather and of stock market fluctuations show that although the models offer far greater predictability than was the case even 20 years ago, there remains a large element of uncertainty in the predictions made. Very occasionally they are spectacularly wrong, but more usually they offer predictions upon which we can rely within certain limits – to carry an umbrella, for example, or to risk moving cash from shares to property investment. The reason that, although models of weather behaviour and of national and world economies have had millions of pounds thrown at them, they retain so much uncertainty is that the phenomena which they attempt to describe contain numbers of chaotic relationships. 'Chaotic' here has a technical meaning, that of a relationship in which infinitesimal changes in the value of an input variable produce enormous differences in outcome variables. Imagine a water molecule in a rain drop falling on the Alps, hitting a rocky ridge. The accident of its trajectory, the influence of winds and any deflection in its flight will determine whether it falls just to one side of the ridge and finds itself travelling in a northerly and westerly direction, becoming part of the Rhine and ending up in the North Sea, or a millimetre away and so southerly and easterly as part of the Danube and into the Black Sea, thousands of kilometres from the outflow of the Rhine. A minute initial variation produces a completely disproportionate outcome.

Before tying this ramble back to professional development, we should note an important difference between our interaction with economics and with meteorology. For the latter, we make virtually no attempt at influence. Apart from some seeding of clouds to encourage them to rain, we generally consider attempts to influence the weather to be the preserve of irrational superstition. The study of economics, in contrast, is largely driven by the belief that if it can be adequately understood then we may be able to take some control over it. If only the operation of the levers of interest rates, taxes, subsidies, and tariffs could be better understood then we might protect ourselves from economic depression, devaluation, and mass unemployment. The reason that we think we can control economics but not the weather is probably less to do with differences in complexity of the models than with the size of the forces involved. The treasury can put up interest rates at the stroke of a pen, but there's not much we can do about sunspots or whorls in the jet stream in the upper atmosphere – even if we know what would be the effects of tinkering with such forces.

Consider now the other end of the series, Newtonian physics. Here we have an ordered world in which chaos plays no part. Given particular values of force, mass, direction, friction, and other well defined and measurable variables operating in ideal conditions, one can predict precisely the velocity and position of a vehicle at any given time. NASA can put a person on the moon, lift them off and deposit them within a square mile area of the Pacific Ocean using Newtonian physics, and they can do it every time, within limits set by the mechanical reliability of the craft. True, to get two satellites to dock in space requires an Einsteinian correction to Newton, but that still uses equations which work.

The conclusion to this rather heavy-handed metaphor is at hand, as the professional development of teachers takes its place in the series as less deterministic than physics, but considerably more deterministic than economics or meteorology. Certainly there are chaotic relationships involved, in the sense outlined above of very small differences in some inputs leading to effectively unpredictable differences in some outcomes. Rosenholtz (1989) uses the idea of 'teaching uncertainty' – uncertainty that arises because the outcomes of teaching are unpredictable because of the variance in students. This has similarity to the notion of chaos introduced here but is rather more specific.

Notwithstanding the acceptance of some uncertainty or chaos, we maintain here that there is also a good degree of predictability. Thus we position ourselves between the rather positivist stance of scholars such as Thomas Guskey (2000, pp 76,77) and the postmodernists such as Andy Hargreaves (1994). The radical post-modern position appears to suggest that since so much of importance in the world is a matter of interpretation, of personal construction, and has no objective reality which can be reliably described by one person to another, then the scientific study of factors which enhance or hinder the effect of an activity such as professional development cannot, in principle, yield useful results. To refute this position requires the establishment of relationships between input variables (in our case, for example, the longevity and intensity of a PD programme) and the desired outcome, reliable change in pedagogy and related changes in students. It is a rejection of the radical post-modern position and a belief in at least the partial predictability of effects from causal factors which must underpin both experimental and correlational research into effective professional development. That such research has produced a range of correlational, if not causal, models (to be explored in the next chapter) which overlap with considerable areas of agreement shows that there is predictability in the system and that the continuing pursuit of better equations, of tighter models with even greater predictive validity, is not a fruitless quest, just as long as we never pretend to ourselves that the system can ever be more than partially determined.

PROCESS-PRODUCT RESEARCH REVISITED

As Guskey (2000) has elaborated, there are a number of levels at which professional development programmes can be evaluated. The most trivial (which Guskey categorises as level 1) form of evaluation is the questionnaire given to participants at the end of an in-service day. "Sadly, the bulk of professional development today is evaluated only at level 1, if at all." (p. 86). At the top level, at least for a professional development programme whose aim is to equip teachers with the skills required to raise their students' general thinking ability, the most compelling evaluation is in terms of gains in student ability and achievement which can be attributed unequivocally to the professional development of the teachers. This is not a simple thing to do, and many have claimed that such process-product research is so difficult that it cannot be done, and should not even be attempted. Richardson, (1994) makes the case that much investigation of teacher development in the past has been very instrumental, treating teachers like objects to be manipulated in a vain search for sets of teacher behaviours which can be relied upon to deliver good student learning. As a reaction to such dubious practices, the trend in classroom research has shifted towards ethnographic studies of classroom ecologies. Here we propose that while ethnographic studies have value for certain purposes, both socio-political and professional voices are quite reasonable in requiring some measure of outcome from investment in staff development, and that process-product research not only can yield useful information, but is the only approach which can in principle provide guidance to teachers and teacher educators on how professional practice might be changed to yield higher student achievement. Firstly, some of the specific criticisms of process-product research should be considered.

Doyle (1977) criticises studies in which specific teacher behaviours are correlated with student outcomes for the idiosyncratic way in which particular behaviours are chosen for study, and the unwarranted assumption of causality underlying the correlation. He compares the process-product paradigm unfavourably with the 'classroom ecology' paradigm:

> "...the purpose of the ecological paradigm ... is to build and verify a coherent explanatory model of how classrooms work, a model that can be used to ask questions and interpret answers about teacher effectiveness" p.176

It is clear that ethnographic studies of classrooms can - at a cost - provide far richer accounts of what happens in classrooms than can simply quantitative studies (see for example, Gardner, 1974; Tobin, Kahle, & Fraser, 1990). But whilst such studies provide rich descriptions, it is less clear how they can lead to prescriptions, that is, to advice to teachers or teacher educators about ways of improving their practice.

Fenstermacher (1979) also makes much of the problem of causality. He exemplifies the point with correlations found between, for example, the use of probing follow-up questions by the teacher and student achievement. He concludes that there is no way of telling from this correlation whether it is the nature of the questions that causes enhanced achievement, or whether higher achieving students provide feedback to teachers which encourages them to use higher level questioning techniques. Such criticism can be met by intervention studies, in which a teacher behaviour postulated as causally related to student achievement is specifically introduced, and changes in student outcomes observed. Fenstermacher's main criticism, however, is that process-product researchers necessarily, and unconsciously, make assumptions about what counts as "good" education. He claims that quantitative researchers are unaware that the products they strive for are no more than culturally determined norms. But how important is such awareness? If teachers, students, parents, university admissions tutors, and employers all agree that test grades are the best measures available of achievement and aptitude then it seems that aiming for higher grades is a perfectly respectable aim for teachers and teacher educators. Evaluation of in-service programmes for teachers whose aims are the development of pedagogy may legitimately look for evidence of increased student performance on measures which have wide popular credibility.

A further problem with process-product research is interaction between particular teacher behaviours and particular learner personalities, learning styles, or context, which makes generalisation of results from individual studies difficult. In an elegant study, Gardner (1974) showed how the use made by different pupils of a given teacher behaviour was mediated by personality, such that the application of a simple process-product model could easily lead to erroneous conclusions. Where a particular teacher characteristic at first sight appeared unrelated to pupil performance, deeper analysis showed that it positively affected pupils of one personality type, and negatively affected pupils of a different personality type.

Brophy & Good (1986) in a thorough review of process-product research recognise all of these problems, and after eliminating studies which fail to meet their rather stringent criteria for acceptability, conclude

> Despite the importance of the subject there has been remarkably little systematic research linking teacher behavior to student achievement. A major reason for this is cost. (p.329)

They mean, of course, the cost of thorough and well designed studies. They find, however, that with more sophisticated observation methods and experimental designs, some reliable relationships began to be established between certain teacher attitudes and behaviours (such as warmth, business-like manner, enthusiasm, organisation, variety, clarity, structuring comments, probing follow-up questions, and focus on academic activities) and students' achievement. They conclude that process-product research is viable, but that it is difficult and requires careful attention to experimental design and interpretation to make its findings valid and usable.

Even if general criticisms of process-product research can be met, there remain two problems particular to professional development which have received less attention in the literature. The first is the dilution effect. A professional development programme can only be one of many influences on a teacher, and a particular teacher can be only one of many influences on the students. The effect of one particular staff development programme may be so diluted in its effect on students as to be undetectable.

The second is the difficulty of isolating sources of failures of an in-service programme. In-service courses are often based on unsupported assumptions about what constitutes effective teaching and learning. The measurability of outcomes associated with such assumed good practice presents a problem. If you are not sure whether or not teaching method X works, in any sense, then evaluation of an in-service programme designed to introduce method X which shows no gain in pupil learning shows either that the in-service programme was poorly delivered, or that method X does not work anyway. There is no way of telling which.

Both of these problems can, in principle, be overcome: by making the staff development programme sufficiently extensive so that its effect is substantial, and by evaluating the methods being advocated separately and establishing that, at least under optimum conditions, they can indeed lead to enhanced student achievement. As part 2 of this book has shown, professional development for cognitive acceleration, as well as applying both of these principles, has used a wide range of methods of evaluation. In the next section we will consider these methods in the light of the general criticisms and problems associated with PD discussed in this chapter.

METHODS OF EVALUATING PROFESSIONAL DEVELOPMENT

> In fact, it is getting harder to find *any* methodologists solidly encamped in one epistemology or the other. More and more "quantitative" methodologists, operating from a logical positivist stance, are using naturalistic and phenomenological approaches to complement tests, surveys, and structured interviews. On the other side, an increasing number of ethnographers and qualitative researchers are using predesigned conceptual frameworks and prestructured instruments, especially when dealing with more than one institution or community. Few logical positivists will now dispute the validity and explanatory importance of subjective data, and few phenomenologists still practice pure hermeneutics – and even those believe that there are generic properties in the way we idiosyncratically "make" rules and common sense. (Miles & Huberman, 1984) p. 20.

Amen to that, we say. It will be clear to any reader of chapters 5 to 9 that in assessing the effects of cognitive acceleration and the professional development which forms an integral part of effective implementation of CA we have used a wide variety of methods from the tightly quantitative to the ethnographic, including en route the type of qualitative approach advocated by Miles & Huberman (1984). In this section we will consider each of these methods critically and, where appropriate, point to other studies which have used similar methods or which point the way to improvement of the methods.

Quantitative – student achievement

The most obvious attempt we have made at process-product research is the quasi-experimental methods described in chapter 5, where immediate cognitive gains and long-term academic achievement of students following cognitive acceleration programmes are compared with controls. Our examples show both the strengths and the weaknesses inherent in such methods. On the positive side, one can conclude reliably that the CA programme has had real and lasting effects on students. The nature and extent of the control groups ensure that the only systematic difference between CA and controls is the CA programme. The size of the sample and the size of the effects offer high levels of confidence that the effect is real, and not an artefact of sampling or chance. As intervention studies these are genuinely experimental, allowing us to impute a clear causal link between the input, the CA programme, and the outcome, student academic growth.

We should mention here a true experimental study by our friends in Finland (Hautamäki et al., 2002). Whereas our studies were 'quasi' experiments (Campbell & Stanley, 1963) since ascription to experimental or control conditions was done on whole class or whole school bases, the Finnish study was a true experiment, a real rarity in the educational research literature. All of the children in one year group in the 20 schools in a small Finnish city were ascribed individually and randomly to one of three conditions: CASE, CAME, or neither. Those students ascribed to CASE travelled by bus every two weeks to a location where they received the *Thinking science* lessons taught by Jorma Kuusela, and likewise with the CAME students. The logistics of this operation must have been formidable, not to mention the strain on Kuusela himself who taught all of the CASE and CAME lessons over and over again for two years. While it is the experimental design which is of particular interest in this chapter, we cannot forbear from giving the results: CASE and CAME children made significantly greater cognitive gains than the controls, and far greater than the national norms. So far, so expected, but what was remarkable was that the control children also made cognitive gains significantly greater than national norms. The researchers propose, plausibly in our opinion, that this was the effect of social construction during the normal classes in the schools. In any class in the relevant year groups in that city on average two-thirds of the students would have received CASE or CAME lessons. One might suppose, then, that the level of questioning, the quality of argument, and the willingness to engage in constructive dialogue would have been higher in all of the classes because of the influence of the two-thirds, and that the one-third benefited directly from this heightened quality of discourse.

What are the flaws in such quantitative approaches? The main one is that one can say with absolute confidence only that "CA" has had the effect. But "CA" is a complex of psychological theory and classroom techniques introduced through a set of printed materials and an extensive professional development programme. Which of these elements is it that is doing the work of cognitive acceleration and academic growth? The straightforward quantitative studies cannot answer this question. In principle it should be possible to test for the gross effect of the PD programme by looking for cognitive gains in schools which have bought the materials and are implementing the programme on their own without any professional development, but identifying such schools and co-opting them to partake in the testing programme would not be easy. There would also be a question about the typicality of such schools: even with the massive support provided by the PD programme schools struggle to take on board the changes in pedagogy characteristic of CA, so any school which was able to do it unaided would probably be quite untypical in the level of enthusiasm and commitment of its science department.

In spite of these problems, it was possible to gain some insight into the effect of the PD from some supplementary analyses of the quantitative data. In one case it was shown that the 'CA effect' actually happened before it might have been expected to, and this effect was attributed to the general changes which had occurred in the teachers, most probably a direct result of the PD. Then, by looking at individual class results, it was possible to demonstrate that virtually every teacher in schools which participated in the PD programme showed cognitive gains greater than control mean gains. This was interpreted as indicating that the PD programme was successful in reaching all teachers in participating schools, whether or not they themselves attended the PD days. It must be admitted, however, that an alternative interpretation is that it is the materials that are doing the work rather than the PD programme.

While such quantitative studies are necessary to establish that CA can indeed have an effect, we need to employ other approaches to try to identify the particular role of the professional development programme in the process.

Quantitative – teacher change

In chapter 6 we described a model-based study of some mediating variables hypothesised as promoting or hindering the implementation of the CA innovation introduced through the CA PD programme. Two broad types of instruments were used: a highly structured interview and a questionnaire yielded data which could be directly quantified; and more open-ended interviews with senior managers which required transcribing and further analysis before any form of quantification could be done. Statements extracted from these interviews were categorised in matrices designed to yield trends in aspects of senior management characteristics related to the school's success in implementing CASE, but we did go a step further than Miles & Huberman (1984) recommend in that we ascribed numerical values to rank orders, plotted relationships and calculated correlations between various factors. Naturally we used categorical rather than parametric correlation coefficients. While the Level of Use scale evaluates the impact of the PD at Guskey's (2000) level 4 ("Participants' use of new knowledge and skills"), other elements of this study tapped level 2 ("Organization support and change").

One might take two opposing, but both critical, views of the procedure we adopted for these studies. On the one hand it could be argued that by reducing the qualitative data to numbers we were both losing important richness and nuance from the data and at the same time imbuing the results with a spuriously 'scientific' look associated with numbers. After all, the data are only as valid as the underlying constructs and the instrumentation intended to tap them allow, and both construct and instrument validity are typically sources of considerable uncertainty. To turn them into numbers is a quantification too far. On the other hand, researchers committed to structural equation modelling (path analysis) not only have no qualms at all about putting numbers on constructs such as 'sense of ownership' or 'level of use', but would criticise our naïve correlational analysis as woefully inadequate for extracting the complexity of the model. A recent example of this genre (Desmione, Porter, Garet, Yoon, & Birman, 2002) demonstrates the power of such modelling to link a wide range of professional development input variables to specified outcomes of pedagogical change, given a large enough sample (they had responses from 430 teachers from 30 schools) and a sufficiently well funded project to handle all of the data (five authors to the paper tells us something here). The older but widely quoted study of Rosenholtz (1989) used LISREL (path analysis software) with constructs such as 'goal consensus' and 'teacher collaboration' tapped by questionnaire responses received from over 1200 teachers to establish a series of important relationships within schools as social organisations.

Caught between the Scylla of qualitative richness and the Charibdis of statistical path analysis, we would argue that the quantification of construct values adds significantly to the clarity with which results can be reported and allows for some sense of the magnitude of effects to be conveyed. Treated with appropriate respect, statistical procedures actually enrich the story being told. In fact our inclination would have been to go the whole hog and use structural equation modelling, had our resources and sample size allowed. Correlational analysis may be a poor cousin to path analysis, but it is readily available to smaller studies and provides a cost-effective way of building hypothetical models linking a number of variables together.

Stories from a whole sample of schools

The study of long-term effects of CASE PD described in chapter 7 relied almost entirely on interviews with key stakeholders in each school. It was aimed largely at Guskey's level 3 ("Organization support and change"). It was a necessary outcome of the question – how permanent are effects of the PD programme after it has finished? – that we should aim to get data from all of the schools in one cohort, in this case 13 schools widely distributed across England. With an eye on the extreme pressure under which teachers work, and the lack of any obvious useful outcome to them of being interviewed for this research, we settled in this study for an interview with just one key informant (usually the head of science) in each school. This was combined with data we already had about selectivity, social environment, and ethnic mix in each of these schools to yield brief 'stories' for each of the schools. We have been careful not to describe these as case studies since they do not accord with the basic tenet of triangulation required before a true case study can be reported with a high degree of validity. Nevertheless, these 'stories' have yielded extremely useful information about some factors which seem to work either for or against an innovatory professional development programme having a long-term effect on a school, its teachers, and students.

Real case studies

In contrast with the school stories described in the last section, the case studies reported by Nicki Landau in chapter 8 are grounded in a large amount of information gained from multiple interviews and informal conversations with the individuals, observations of their lessons and of PD sessions, and conversations with other key informants. All of these data were recorded, sifted, sorted, and cross-referenced so that all assertions in the case studies can be shown to be rooted in evidence, generally from more than one source. These case studies use techniques and yield data at Guskey levels 3 and 4.

This study has added enormously to our understanding of the detail of the processes of implementation of an innovation in a school, especially concerning the interplay of a teacher's personal beliefs and attitudes with the school environment: the social atmosphere, power games, departmental attitude to innovation, and the headteacher's extent and quality of support and interest. It provides a deep filling-in of detail to some parts of the skeleton of understanding of relationships between key variables which has been constructed by the other studies described in chapters 5 to 9. It is important to note that what has been reported in chapter 8 is only a small sample of the total study which included many more teachers, sampled with as much attention to representativeness as can be given in an in-depth qualitative investigation. We should note that Nicki's research is very different from studies sometimes reported which arise from a researcher's detailed observation of a single teacher. Such studies are so threatened by issues of idiosyncrasy that it is rarely, if ever, possible to draw any general conclusions from them.

We believe that these case studies illustrate well the power of an in-depth ethnographic approach but we need to recognise also the cost: one person's virtual full-time occupation for three years plus many months of work subsequently on analysis and construction of the case studies.

An alternative use of multiple sources

Gwen and John Hewitt's work reported in chapter 9 illustrates a different and complementary method of using extensive and multiple sources of data from interviews, observations, and written reports. As with the case studies, this investigation yields data at Guskey's levels 3 and 4. It is relevant that this was work commissioned by the local education authority specifically as an *evaluation*. That is, the Authority was implicitly concerned to know whether it was getting value for money from the cognitive acceleration programme in Year 1, and particularly from the professional development systems – traditional and teacher-tutor based – which were in place. The researchers also had in mind the very practical outcome of improving the quality of the PD that was being offered, in other words saw their evaluation as being essentially formative. This purpose to some extent dictates the way in which the data is processed. A very open-ended question of the type which drives Nicki Landau's work, which is of the form "What goes on with teachers as they participate in a PD programme?" lends itself naturally to ethnographic methods which have case studies as one important form of outcome. On the other hand more focussed questions of the sort addressed by the Hewitts – "Is CA@KS1 effective in schools?, are the PD programmes effective?, how can they be improved?" - lead naturally to more counting of response categories, reporting of percentages satisfied or unsatisfied (for example), with the raw numbers enriched with extensive illustrative quotations. We suggest that it is the difference in purposes of the two studies which leads to the different forms of treatment of data sets which are of very similar form. Furthermore, we would contend that *both* methods of dealing with the data, of drawing conclusions, and of presenting results are equally valuable, but for slightly different audiences. (Guskey, 2000) refers to Worthen & Sanders (1987) for a discussion on the differences between research and evaluation.

CONCLUSION

In this chapter we have argued that although there are chaotic relationships which intervene between input variables and outcome variables in the study of the professional development of teachers which preclude the possibility of a high level of determinacy, it is still possible to establish some reliable relationships. We have looked at a wide variety of methods which have been used in parallel with one another, from the quantitative establishment of a causal link between cognitive acceleration and desirable student-level outcomes, through to an entirely qualitative ethnographic study which tells us a lot about the nature of what goes on along that chain from PD to student outcomes. We are fortunate that the envelope of cognitive acceleration activities is large enough to permit many different methods of investigation to be used and to complement one another in building an overall picture of the effectiveness of its professional development programmes. The conclusion to this chapter is neither more nor less than reinforcement of the quotation from Miles and Huberman with which we started the section on methods of researching professional development (p. 143): it is that no singular method of investigation will ever be sufficient. A multi-method approach to research into anything as complex as the professional development of teachers is the only one which will confine the chaos factor to a limited area in the research space, and allow apparent relationships to be revealed with a number of different spotlights.

In the next chapter we will consider each of those relationships which are highlighted in our work, and draw on the work of others to reinforce or challenge our conclusions and help us to establish validity.

11. ELABORATING THE MODEL

In this chapter we are going to draw together the main conclusions which arise from our own research and experience and from the work of others concerning the impact of various factors on the effectiveness of professional development. We will build a model which incorporates these variables and which is based on more than supposition – that is, which has real empirical support. But before we embark on this elaboration, a word is in order about such models in general.

We proposed in chapter 10 that the relationship between educational, environmental, and social inputs (independent variables) and school outcomes (dependent variables) includes a chaos factor which precludes the possibility even in ideal conditions of a high degree of determinism. As Thiessen (1992) says:

> The classroom is more than a dependent variable patiently waiting to obstruct or welcome the passage of independent variables, such as enquiry, into its midst p.88

Nevertheless, we have argued that a level of predictability is attainable, and that this justifies the process of building models on the basis of evidence, however incomplete, since the models offer an opportunity for the generation of hypotheses to be explored in further research. Furthermore, as we will argue in the next chapter, policy-makers need guides to action now rather than when 'more research needed' has been completed, and an imperfect model which has some substantial empirical justification is a lot better to work on than no model at all.

We propose to set the model within a general theoretical framework in which the variables can be conceptualised. This framework is outlined in the next section.

THEORETICAL FRAMEWORK: THREE STRANDS

Wilson & Berne (1999) conclude their review of research on professional development thus:

> Our review of the literature leads us to conclude that ... few such projects had yet completed analyses of what professional knowledge was acquired... Fewer still had explicated their theories of how teachers learned and designed research to test those theories. (p. 204).

Fullan (1995) also suggests that

> Professional development of teachers has a poor track record because it lacks a theoretical base and coherent focus p.253

These may be a rather harsh judgements. We suggest that there are at least three strands of theorising which have proved fruitful in understanding the process of professional change of teachers. The first is the application of general theories of conceptual or attitude change to the beliefs and behaviours of teachers; the second is the notion of the reflective practitioner; and the third is an emerging body of scholarship which characterises teachers' procedural knowledge as 'intuitive'. Let us consider each of these strands in turn.

Conceptual Change

Borko & Puttnam (1995) offer us a cognitive-psychological perspective on professional development in which change in practice is associated with changes in the inner mental workings of teachers and their constructions of new understandings of the process of learning. An example of approaching professional development in the context of conceptual change is provided by Mevarech (1995) who discusses the role of teacher's prior conceptions of the nature of learning and describes the U shaped learning curve which teachers encounter when trying to replace one skill, and the epistemology on which it is based, with another. Bell & Gilbert (1996) also approach the issue of the professional development of teachers from a constructivist perspective, showing how teachers need to interrogate their own current constructs of teaching and learning before they are ready to re-construct new beliefs. The value of this conceptualisation of teacher change is that it can draw on the extensive parallel literature on conceptual change and attitude change in students. It leads us to focus on teachers' prior conceptions and to recognise that we are unlikely to bring about change in practice unless we face up to and, if necessary, challenge teachers' deep-rooted beliefs about the nature of knowledge transmission. It suggests the idea that teachers should be included in the process of needs analysis and programme design (Joyce & Showers, 1982), to which we will return. It also indicates that such change is likely to be a slow and difficult process, and that real change in practice will not arise from short programmes of instruction, especially when those programmes take place in a centre removed from the teacher's own classroom.

Note that in focussing on the need to tackle fundamental concepts and attitudes, we are not necessarily prescribing that this is the first thing that must happen, before change in teaching practice can occur. Indeed Guskey (1986) has argued persuasively that changes in teachers' beliefs and attitudes may well follow the change in perceived pupil responses which come about from changed teaching practice. Nevertheless, whether they are a precursor or a consequence, such deep-seated changes are necessary for permanent effects on teacher practice.

Other researchers have reinforced this view and focused on the importance of the individual needs, the motivation and the expectations of teachers. They all emphasise that the beliefs and value systems of teachers need to be, or become, congruent with the message of the INSET provider ('value congruence').

Reflection on Practice

The idea of the teacher as a reflective practitioner has had a long and respectable history in the literature. For example, Baird, Fensham, Gunstone, & White (1991)

have shown, particularly with respect to the professional development of science
teachers, the central role that reflection – both on classroom practice and on the
phenomena of science teaching and learning – has in the pedagogical development
process of both pre-service student teachers and experienced teachers in inservice
courses. More recently Cooper & Boyd (1999) have described a scheme of peer-
and group- oriented reflection on practice developed amongst teachers in a New
York City school district which provided a systemic self-help strategy for the long
term maintenance of innovative methods in classrooms. Fullan (2001) also makes
much of this project in his investigation into effective educational leadership.

The notion of the reflective practitioner arising from such studies guides practice
in professional development by highlighting the importance of allowing teachers
time to discuss their current practice and their attempts at changing it during the
course of a programme. This may be done through diaries or other forms of logs, or
orally at 'feedback' sessions with colleagues and course leaders. In fact the use of
the term 'feedback' for this process may be somewhat misleading. Feedback may
mean providing an account of the success or otherwise of a new procedure or
activity so that the course developer is able to modify the material. On the other
hand, in feedback *as reflection* it is specifically the participating teachers who
benefit through putting their experiences and associated feelings into words and
discussing them with peers. Experienced course organisers know that such sessions
can become self-sustaining and that many of the ostensible 'questions' that arise are
in fact answered by participants themselves or even serve a rhetorical function:
asking the question provides its own answer. We might ask then whether such self-
driven reflection is by itself sufficient to create change in practice - in other words,
is there further input required from course leaders who have deeper academic
insights into either or both of subject content matter or of learning theory? A
number of scholars (Avgitidou, 1997; Calderhead, 1993; Korthagen & Kessels,
1999) argue from various perspectives that the subject matter of reflection must
include, *inter alia*, a richer understanding of the psychological principles of teaching
and learning. This gives the leaders of a course of professional development a
responsibility to support participant teachers in building such an understanding for
themselves, and indicates that reflection alone will generally be inadequate to
generate change without some input from a more experienced and theory-driven
perspective.

Intuitive Knowledge in Teaching Practice

Turning now to the intuitive nature of much of the procedural knowledge of
teachers, it is important not to confuse the ideas of 'intuitive' and 'instinctive'. The
latter implies something in-built, perhaps a personality factor over which no normal
professional development course could be expected to have much influence.
'Intuitive', on the other hand, implies a behaviour which occurs without explicit
cognition at the moment at which it arises. The basis of the behaviour remains in the
unconscious. The term 'implicit knowledge' is used for this type of unconscious

understanding which gives rise to intuitive behaviour (Tomlinson, 1998). Intuition is how, as teachers, we react almost instantaneously to situations as they arise in the complex social environment of the classroom. It would be practically impossible for a teacher to proceed in such a situation entirely on the basis of rational and conscious decision-making or problem-solving. The 'professional' response in such situations depends much on intuition, a process well described by Brown & Coles (2000) through their collaboration between a researcher and a practising teacher. The important point here is that this intuitive behaviour is based on our implicit knowledge, and that knowledge is based on previous situations and on the constructs we have built on such experience but not necessarily externalised or made conscious.

> ...it is now well accepted that expert practitioners posses a complex personal knowledge base which they draw upon intuitively. This knowledge base is acquired through training and experience but individuals may not be able to articulate why they do what they do. (McMahon, 2000 p. 138)

Such implicit knowledge may be an influence for good or for ill in the direction it proposes for action. Implicit knowledge can be derived from working in a traditional context rooted in an authoritarian view of teacher-student relationships and based on a simple transmission epistemology. On the other hand, it may be derived from a combination of a personal philosophy of guided democracy, with some experience of the process of constructivism, and the observation of colleagues who have shown how all students can be encouraged to contribute to the construction of their own understandings. This relates to the 'professional' strand in Bell & Gilbert's (1996) three-part model of the professional development of science teachers. The challenge for providers of professional development programmes is to devise programmes which shape teachers' intuitive understandings such that they improve the quality of daily classroom reactions. Such a process is necessarily slow and cannot follow a simple mechanistic path.

It must be emphasised that the three strands of thought on the nature of professional development outlined above (concept change, reflection, intuition) are not seen as alternatives. On the contrary, they intertwine and feed into one another. What is an effective way of inducing a process of conceptual change in a teacher? Why, to encourage reflection. And what is the basis of the intuitive knowledge which guides action? It is the underlying conceptions and attitudes of the individual. Guided reflection assists the process of conceptual change, and conceptual change re-structures the intuitive knowledge upon which teaching practice rests. In his seminal work on professional development, Schön (1987) shows how reflection is an essential part of the process by which teachers incorporate the perceived needs of a situation within their own system of beliefs, and this is all part of the development of their 'professional artistry'. This is a good description of practice arising from implicit understandings.

KEY VARIABLES

Before proposing a model of the factors which impact on the professional development of teachers we need to identify the main variables. It will be taken for granted that the outcome (dependent) variables are (1) changed pedagogical practice in the classroom and (2) consequent beneficial changes in students, related to their intellect, to their achievement, to their motivation, or to other characteristics for which the educational system must take some responsibility. We will focus, therefore, on the input (independent) variables and on mediating variables and where appropriate relate them to the theoretical framework outlined above. We will consider the variables under three headings: the nature of the innovation being introduced; the nature of the delivery system; and the nature of the environment into which it is being introduced. For each variable we will say something of its nature and describe and justify our best estimate of the relationship between that variable and effectiveness of professional development.

1. Nature of the innovation

Fullan & Stiegelbauer (1991) emphasise the pointlessness of organising professional development for an innovation which is not itself worthwhile or of established quality. They attribute failure of the post-Sputnik reforms in science education in the United States to the fact that the innovations were driven by politicians and had not been established as educationally sound. (One might note in passing that the story of the UK equivalent, the Nuffield science programmes of the 1960s, is rather different (Waring, 1979). Here the reforms were driven by teachers, but they were teachers in private and selective schools and so their educational value was limited to that sector of the school population). Fullan also quotes Cuban's (1988) view that first order changes (doing what we do, but doing it better) are far easier to implement than second order changes (doing something different). We have shown that cognitive acceleration is a worthwhile educational innovation, and we would claim that it is a second order change for most of the teachers we work with. So, difficult but not impossible.

The general message is clear: a prerequisite of effective professional development is that one must have good reason to believe that the change being introduced is itself of value. Looking at the nature of the innovation in a little more detail, we would highlight two characteristics. With respect to the first strand of the theoretical framework outlined above, it seems clear that the chances of bringing about changes in teachers' conceptualisation of the teaching-learning process must be far higher with an innovation which makes good educational sense and for which there is some evidence of effect.

1a. Theory base

Fullan & Stiegelbauer (1991) and Miles & Huberman (1984) are clear that successful innovations must have some sort of conceptual framework and theoretical foundation. We have shown that teachers actually welcome the opportunity to share in the reasons why they are being asked to change their practice. Not to offer them some insight into the rationale underlying an innovation is to treat teachers as technicians rather than as professionals. As we have said earlier, the power of the professional lies in the ability to be flexible, to change details of the practice in changing circumstances, because the practice is rooted in understanding of purpose and is far deeper than the blind following of given procedures. A theory base aids the process of conceptual change and, in building changes in pedagogy which become intuitive, provides a touchstone for new practices.

1b Evidence for effects

The innovation being introduced by the professional development programme must be one for which there is evidence, or at the very least good reason for believing, that it will in fact have a positive effect on teaching and learning. This seems so obvious, and yet schools are continually being bombarded with claims from one guru or another for the wonderful effects of their particular brand of snake oil, for which there is actually no evidence of any effect at all.

1c. Generative activities

Curriculum activities alone, presented in however a sophisticated mixture of print, visual, and software resources, will never bring about educational change. We have continually emphasised that real change lies not in materials but in people, in the teachers. At the same time, we need to recognise the need for supportive materials of adequate quality and accessibility. One is reminded of Rabbi Lionel Blue's story about the devout man who pleads with God to let him to win the lottery since his family is hungry and he cannot find work. After praying fervently every Saturday for three weeks in a row, there is suddenly a crash of thunder and the voice of God booms down: "Maurice", says God, "I hear you, but you have to meet me half way – at least buy a lottery ticket". It is asking a great deal to expect teachers to change their practice solely in response to presentations of theory and strategies and participation in activities during inservice days or classroom visits by tutors. New practice needs all of those things but it also needs to be supported with materials to which the teacher can refer when she or he is on their own. Learning to change ones teaching practice is both an academic and a practical process and successful change requires a continual iterative process between references to the academic, written, materials and the trialling of new pedagogies both alone with reflection and with the support of peers or a tutor. Such materials support both reflection and the building of practice into the intuitive mode.

Key characteristics of useful support materials are that they should be accessible – clear with well-constructed ways of finding ones way about – and that they should be generative of the type of pedagogical innovation being introduced. In the case of our cognitive acceleration work, the activities were developed specifically to maximise opportunities for cognitive conflict, social construction, and metacognition. Were the aim of an innovation to be, to take a different example, to improve students' construction of science concepts, then support materials would need to be generative of conceptual constructivism.

2. Elements in the PD programme provision

There is universal condemnation in the research literature on professional development for the one-shot 'INSET day' as a method of bringing about any real change in teaching practice. Perhaps the only exception to this rule is the introduction of a very specific technical skill, such as the use of new piece of software. But if one shot does not work, how many shots, and of what calibre, are required? We will look first at the quantitative question.

2a. Longevity and intensity

Fullan and Stiegelbauer (1991, chapter 4) think 2 years is a minimum for real change to occur and Joyce & Weil (1986) believes that a new pedagogic skill requires 30 hours of practice before it is perfected. With our 'original flavour' CASE PD we meet both of these criteria since the PD course itself is delivered over two school years (see chapters 3 and 4) and the teachers involved implement a minimum of 30 one-hour teaching activities which incorporate the pedagogy of cognitive acceleration. We have much anecdotal evidence that real change in intuitive practice frequently does not start to occur until well into the second year of this programme. With our newer work in Years 1 and 3, however, alternative models have been attempted. Both of these have been projects to develop CA materials and methods for just one year. In the case of the Y1 work, the application of the methods were so new to us as well as to the teachers that we made the decision to concentrate our resources on one year to maximise the research evidence we could extract. To have attempted to work in Y2 at the same time would have doubled the number of activities to be developed, the number of teachers involved, and all of the associated observations and data collection. So here we broke the 2-year rule, but the intensity was high. Each teacher taught a half-hour CA activity every day of the week for 30 weeks, so they had something like 75 hours of practice with the method over the year. Our evidence (see chapter 9) is unequivocal that this programme did work at the level of changing teaching practice, and we also have substantial evidence for immediate effects on children.

With the Year 3 work, however, we broke both of the criterial rules. The PD programme extended over one year, and the CA activities were used with the whole class only fifteen times over the year. This work is too new to have generated a separate chapter for this book, as data is still being collected as the book nears completion. But we can report here that although there is good evidence of change in teaching practice, there has been no significant effect on the children's cognitive development or gains in science achievement. That is, in terms of Joyce's stringent requirements that a PD programme must show evidence of effects in students to be judged effective, this programme has not been effective. Our current intention is to extend the work into Year 4, thus making it a two-year / 30 hour programme.

This experience suggests that although there can be some trade-off between the intensity and the longevity of a programme, the general guidelines of 2 years / 30 hours are sound. That this is a minimum requirement is entirely consistent with the theoretical framework described above, since conceptual change is well-known to be slow, and the building of new skills into intuitive practice must require plenty of practice aided by opportunities for reflection.

Under this heading of length and intensity, there is also an issue of 'lead-in time'. It turns out that the planning time for introducing an innovation is somewhat critical. Evaluation of our first cohort of CASE training (chapter 6) showed that offering the first INSET days of a two year programme at the beginning of the school year in which implementation was to start caused an element of panic as no reflection time was allowed between getting back to school and starting the programme. We immediately changed the CA PD programme so that the introductory days were held in July, at the end of the year previous to that in which CASE teaching was to start. This provided practical time for obtaining resources and sharing information within the department, but more importantly a gestation period in which the CASE co-ordinator and other members of the department could consciously reflect on and unconsciously process the information they had gained. On the other hand, the lead-in time must not be prolonged unduly. Guskey (2000) recounts tales of implementations which have become so bogged down in pre-planning meetings and the sharing of information to take everyone's view into account that the innovation never gets off the ground at all. There has to be an element of 'Oh let's dive in and see how it goes.'

2b. PD teaching methods model the target classroom methods.

Nothing is less convincing, or more ironic, than a formal lecture on the benefits of constructivist teaching as part of a professional development course. It seems obvious that we are unlikely to encourage teachers to use active methods in their classroom by delivering to them a monologue and expecting them to take notes. This is another of those paragraphs which would hardly seem worth writing, had we not ourselves frequently experienced such mis-matches between message and delivery method ourselves and heard many tales from teachers of similar experiences. So, yes, we will spell it out: if you want to promote the use of cognitive conflict, then present your teacher audience with some cognitive conflict at their own level. Anne Robertson has a quiver full of problems which really make teachers think, after which she asks them not just metacognitive questions – ' how did you solve that?', but also meta-emotive questions: 'how did you feel when faced with the problem?' Responses such as 'panicky' and 'I'd hate to have to do that if I wasn't surrounded by sympathetic colleagues' open the way to a discussion about how their students are likely to feel in similar situations. In the same way we have activities for teachers which can only be solved by collaboration with colleagues, which force social construction on all participants. Only in such ways can the notions of 'cognitive conflict', 'social construction' and 'metacognition' gain some sort of reality in the INSET context, before they are practiced in the classroom. This is the first step in the process of teacher conceptual change, reinforced by reflection induced by metacognitive questions. At this point the methods remain explicit and largely external. Only with extensive practice do they become intuitive.

2c. Coaching / Reflection

Coaching is now known to be a critical process in assisting teachers to transfer approaches they have studied in inservice sessions back to their own classroom settings. From a meta-analysis of nearly 200 studies of the effect of professional development, Joyce & Showers (1988) concluded that of all the features which are normally incorporated into professional development programmes (such as the provision of information, demonstration by trainers, opportunities to practice the new method, provision of feedback, and coaching of participants in their own schools), it was coaching which proved to be an essential ingredient when the outcome measurement was student change. This finding is consistent with the decay in effectiveness of PD we noted in the Indonesian PKG project when coaching was withdrawn for reasons of cost (chapter 2).

It is not difficult to relate these empirical findings to our theoretical framework of three strands (conceptual change, reflection, intuition). It has already been emphasised that each of these processes is itself necessarily slow and difficult, but coaching has a contribution to make to all. Firstly, the expert coach can provide a mirror to assist the teacher reflect on her or his current practice. A knowledgeable and sensitive observer can feed back to the teacher detail of actual practice of which the teacher (because the practice is largely intuitive) may not themselves be aware. This detail in turn provides the raw material for reflection. Either at the time of the lesson or subsequently, either alone or with others, the teacher may then be encouraged to examine her own assumptions, beliefs, and concepts concerning teaching and learning which give rise to her current practices. A series of such coaching sessions, which may include also occasions when the coach models particular techniques (such as open questioning, wait time, generating activity from all students), promotes a process of conceptual change in the teacher, and the implicit knowledge thus modified gradually changes her behaviour. Initially, the new practices may be at a conscious, non-automated level, when they may be expected to be somewhat stilted and laboured, but as they become automated and part of the teacher's intuitive practice so they become fluent and more effective. This process explains the U shaped learning curve described by Mevarech (1995) referred to above, and accords well with our own experience of running long-term professional development programmes.

The central role of coaching in effective professional development is therefore well-established, but we still need to explore some aspects of coaching which aid or hinder its effectiveness. Garmston (1987) and Garmston, Linder, & Whitaker (1993) describes 4 types of coaching:
- Technical coaching is aimed to help teachers transfer skills from inservice courses to their classroom. It generally follows staff development workshops and involves consultants and teachers.
- Peer coaching is aimed to refine teaching practices, increase collegiality, and encourage reflection. Pairs of teachers work together and the focus of the coaching is determined by the observed teacher. (We will return to the collegiality issue later)
- Cognitive coaching is aimed to explore teachers' thinking. Typically it involves a pre-conference, observation, and a post-conference. Coaches need to be trained to facilitate the thinking of the teacher and to address their implicit beliefs about teaching and learning.
- Challenge coaching is used in response to a persistent problem, typically in a department or other team. It involves peer observation targeted at specific problem, which when resolved is used by all in team.

Of these categories, it is technical and cognitive coaching which has been the predominant mode of our own work. Challenge coaching is for a different purpose and while we have made real efforts to encourage peer coaching, the pressures of time under which teachers work is such that there has been very little implementation of peer coaching on a regular basis within our project.

Whatever type of coaching is involved, it is important to clarify and establish the role of the tutors (coaches). In educational systems such as those in Indonesia and the UK where there is a strong culture of inspection and appraisal, teachers are quite naturally apprehensive about the role of a strange adult in their class. So the first sub-condition is that tutors must be reassuring, collegial, and supportive. In PKG (chapter 2) and in the Let's Think! work (chapter 9) this condition was met by appointing tutors chosen from amongst their peers for their teaching skill, but not giving them a permanent inspectorial role. These tutors normally maintained their regular teaching job but are given some time to visit other teachers. In CASE at KS3 the problem is met by the way that tutors interact in the classes they 'observe'. Usually they participate in the lesson, maybe simply by assisting with group work or by playing a more active role in delivering the lesson. They present themselves as team-teachers, possibly more experienced and skilled in the specifics of CASE methods, but in other respects little different from the 'observed' teacher.

Secondly, the skill of observing and coaching another teacher in a manner which is supportive and useful is one that tutors have to learn. Exactly what are you looking for when you observe? How do you record important observations? How do you feed them back to the teacher in a way that is not threatening but likely to be productive of change? Within CASE, these are questions continually raised in tutor forum meetings held two or three times a year to share experiences and consider new developments. In work with local authorities, where we expect to train tutors as well as teachers in order to establish sustainable systems (see chapter 9 and the section on Authority-based schemes in chapter 4) the programme includes opportunities to coach coaches.

Finally there is the matter of actually being something of an expert. The tutor is supposed to have something to offer to the teacher. That does not mean having the answers to all possible problems that might arise, but it does mean having considerable experience, and being able to say "yes I've been there, and this is what I did, you might want to try ....". Even peer coaching often involves a hierarchical relationship between the coach and the teacher. A number of authors (Yarrow & Millwater, 1997) have said that it is the power relationships which are inherent in these models that create the problems often encountered when the coaching involves giving feedback. Andy Hargreaves (1992) also warns of the difficulty of peer coaching, and emphasises rather the informal contacts between teachers in a school, which amount to the learning (or otherwise) culture of the school. Partly to avoid these difficulties, some researchers (Joyce & Showers, 1995; Showers & Joyce, 1996) have suggested that the feedback element may even be omitted altogether. This model of coaching sees teachers working collaboratively to solve the problems and questions that arise during the implementation of the curriculum innovation. They report that this collaborative team approach to coaching, without any feedback, does not adversely effect the implementation of the innovation or pupil growth, although they note that even in these situations peers easily slip into a supervisory role in which they give evaluative comments. There is a suggestion also that when teachers coach other teachers the emphasis is likely to be on practical rather than theoretical aspects. Our own experience is that the fear of overly hierarchical or supervisory coaching is more often felt by the coaches than by the teachers being coached. Teachers welcome the advice of colleagues more experienced in particular skills, while new coaches are often nervous of giving clear advice.

Whether coaches are university or LEA-based tutors who come in to the school, or whether they are peers within the school, there is always a significant issue of the cost of coaching, and we will need to consider this further in the final chapter.

3. Environment in which change is engendered

What are the factors which cause an innovation to fade away, to run into the sand, and what are the factors that make an innovation take root and make a permanent change to educational practice? The evidence is overwhelming for the importance of the school environment (or 'school culture') in providing the conditions in which innovation may flourish and be sustained. Notwithstanding the main focus of this book on the teacher her- or him-self, it is clear from the literature and from our experience that teachers rarely if ever are able to make real changes in their pedagogy unless the school environment in which they find themselves is, at the very least, tolerant of innovation. "No man is an island" (Donne, 1571-1613). The case studies reported in chapter 8 offer clear and specific examples of just how the culture of a school presses on individual teachers and either supports or inhibits their attempts to make changes. There are two main aspects to this: one is the presence or absence of collegial support and the opportunity to share experiences informally but frequently; and the other is more related to the extent to which the innovation is embedded in the management structure of the school. In this section will look at these main factors as well as some less salient ones.

3a. Collegiality
Stoll & Fink (1996) list collegiality as one of 10 features of a positive school culture (which include also shared goals and responsibility for success, continuous improvement, lifelong learning, risk taking, support, mutual respect, openness, celebration and humour) but under collegiality they note that:

> this much used but complex concept involves mutual sharing and assistance, an orientation towards the school as a whole, and is spontaneous, voluntary, development-oriented, unscheduled, and unpredictable pp93/94

We discovered early in the CASE project (chapters 3 and 4) that teachers who are trying to change their practice find it extremely difficult to be 'different' from their colleagues in the same school. Schools which were most successful in taking on the innovation were ones in which there was much communication between teachers in the department about the new methods. No one individual, however well motivated and energised, can maintain a new method of teaching if she or he feels isolated. McLaughlin (1994) quoted by Fullan (1995) reported that

> as we looked across our sites at teachers who report a high sense of efficacy, who feel successful with today's students, we noticed that while these teachers differ along a number of dimensions ...all shared this one characteristic: membership in some kind of strong professional community

Fullan & Stiegelbauer (1991) say:

> Within the school, collegiality among teachers, as measured by the frequency of communication, mutual support, help, etc., was a strong indicator of implementation success. Virtually every research study on the topic has found this to be the case. pp 131/132

and Huberman (1995) suggests that

> ... a heterogeneous group reflecting, conversing, debating, and experimenting together is a powerful device for intimacy, mutuality, mastery, and, in the best cases, a re-socialisation or re-bonding to their professional guild p. 218

Desmione et al. (2002) provide strong quantitative evidence for the positive effects of collective participation in a PD programme. Goddard, Hoy, & Hoy (2000) have built the notion of 'collective teacher efficacy' on the basis of Bandura's (1997) concept of individual self-efficacy. They developed a scale of collective self-efficacy and in a study of 47 elementary schools were able to show a clear relationship between collective self-efficacy and the reading and mathematics scores of the students. In another large scale study of elementary teachers in their schools

Rosenholtz (1989) establishes that there is a clear association between teacher collaboration and greater 'certainty about a technical culture and their instructional practice' (p. 46). She finds that

> Teachers who share their ideas, who unabashedly offer and solicit advice and assistance, and who interact substantively with a greater number of colleagues, expand their pedagogical options and minimize their uncertainty p. 111

Joyce (1991) lists collegiality is the first of his 'Doors to school improvement'. All of this experience is entirely consistent with our own data on the importance of communication, reported in chapter 6, and relates closely to the importance which Bell & Gilbert (1996) attribute to the social dimension of professional development.

In terms of our three-strand theoretical framework, it seems clear that teachers' concepts about the nature of teaching and learning will be heavily influenced by the type of 'mutual sharing and assistance' they encounter, and that in a positively collegial atmosphere there will be safe opportunities for reflection and so an increased chance of building new strategies into intuitive practice.

All of this emphasis on the value – nay the necessity – of teachers in a department forming a Community of Practice (Wenger, 1998) leaves open the question of just what this collegiality looks like on the ground. In a simply descriptive manner one can describe a scale from teachers having virtually no professional conversations with one another, through informal chats about the innovation in the corridor or over coffee, to the situation where one or two members of the department have responsibility for overseeing the implementation, and can act as sounding-boards for the others as they try out novel approaches. Better again is the addition of regularly scheduled meetings devoted to assessing progress in implementing the innovation and best of all is some form of peer-coaching. This will lead us into the next section on the role of the headteacher in creating space and the atmosphere for such strategies, but first we should add a note about what might be called the 'unit of collegiality'.

In secondary schools the department is the natural unit of collegiality for a subject-based innovation, while the Year team may form such a unit for a pastoral and age-related innovation. The senior management team may also form a collegial unit for policy and leadership issues. In primary schools the unit of collegiality is more difficult to identify, especially in smaller schools where there is no more than one class in each year group. Rosenholtz (1989) found that in isolated settings teachers' opportunities for growth are limited almost entirely to trial-and-error learning and our own experience has been that the Year 1 teachers we have worked with have often felt very isolated as they struggle to introduce cognitive acceleration methods. For them the INSET days or twilight sessions are a wonderful opportunities for sharing experiences, and thus they form their units of collegiality across schools rather than within them. In dense city areas where we have worked such inter-school sessions – in PD centre or pub – are relatively easy to organise but in rural areas they may not be. Texting on mobile phones and email groups are often offered as solutions, but even if the technology becomes still easier to use and more generally accessible it can never replace the experience of personal meetings.

Note also that units of collegiality within schools form subcultures which may be productive (as in the case of a science department that happily shares both professional and social experiences) or may be 'Balkan' (Stoll & Fink, 1996). Andy Hargreaves (1992 p.223) also refers to balkanisation as the formation of groups of like-minded teachers within a school who can be either generative of change or a carping and disruptive influence. The case studies of chapter 8 illustrate the nature and influence of such subcultures.

3b Senior Management Team

Joyce et al. (1999) place much emphasis on the necessity of effective leadership for the implementation of any educational innovation, as do Fullan (2001); Mortimore et al., (1988), and most other writers on the subject of effective schools. Our experience and evidence fully support the importance of the headteacher for successful implementation of cognitive acceleration in a school, and we would point to two particular aspects where the headteacher's role is necessary, without which the implementation is unlikely to be maintained. The first is in recognising the time required for in-school professional development, and the second is in building the innovation into the structure of school, or at least of the department. These correspond roughly to two of the key features which Fullan & Stiegelbauer (1991) report as essential if an innovation is to become institutionalised: the commitment of the Principal (headteacher), and the incorporation of structural changes into school and classroom policy. We will consider each of these in turn.

All of the strategies described in the last section for maximising productive collegiality depend critically on recognition by managers in the school – typically the headteacher and the head of department – that investment in time for sharing amongst teachers is at least as important as is time for inservice training provided by outsiders. Rosenholtz's (1989) study referred to earlier suggests that the origin of a collaborative atmosphere lies with the headteacher. We have from time to time been quite surprised to find that a headteacher who is prepared to find a significant sum of money for the CA PD programme then baulks at creating the time for teachers to meet together to share experience and to develop their practice collaboratively within the department. This occasional headteacher seems to act as if paying the money was all that was required for magic to follow. The best PD programme in the world will have no deep-seated effect on practice if there is no active support mechanism for teachers introducing new methods, to ensure that the hard work involved in high-quality teaching is recognised, and to establish methods of sharing practice.

The second aspect, highlighted as a common factor in the failing and struggling CASE schools (chapter 7), was the absence of any structural sustainability built into the school. It is the responsibility of senior management in the school to provide systems built into the school management and culture which ensure that a method or approach which has been introduced and which is still considered positively is actually maintained. Taking CASE as an example, practical signals that it has been adopted into the structure include requests from the headteacher for updates on the implementation, attention by management to timetabling requirements of CASE, the inclusion of CASE in departmental policy documents and development plans, the 'showing off' of CASE to inspectors, governors, and parents, and the mention of CASE in advertisements for new teachers. Without the establishment of such sustaining structures, efforts put into inservice work are in danger of being lost when one or two key teachers leave the school.

There is a further feature of the school environment which we will put in this section on senior management, although it is shared. That is what Mortimore et al. (1988) and Stoll & Fink (1996) refer to as 'shared goals' and we have called 'unity of vision'. We investigated in particular the motivation of headteachers and of heads of department and/or CASE co-ordinators in the same school, and characterised them on a scale from 'compatible', through 'independent' to 'incompatible'. There was a strong relationship between compatibility and implementation of the innovation.

We turn now to two variables within the school environment which are perhaps less tangible or amenable to control.

3c. Ownership

In chapter 7 we described how we conceptualised the idea of 'sense of ownership' and showed some evidence that it was related to teachers' acceptance of the underlying theory of CASE, to a motivation of the headteacher and head of department towards student thinking and staff development, and to the provision of formal meeting time to discuss the implementation. We are thus at one with much of the literature in believing that engendering a sense of ownership in teachers for an innovation is important. We may not, however, be in such close agreement about how one best engenders such a feeling. In writings on professional development there is a persistent argument – implicit if not 'hot' – between the 'top-downers' and the 'bottom-uppers', with the latter probably occupying the most fashionable position. Thiessen (1992), for example, in discussing classroom based teacher development, puts the centre of gravity on the teacher and his or her perceived needs. On this view incomers (consultants etc) should primarily respond to these perceived needs, rather than attempt to impose their own agenda. Stoll & Fink (1996) say that

> "A high proportion of school effectiveness efforts worldwide have collapsed because of resistance to the imposition of change" p. 61

The implication is that teachers who feel imposed upon are not in the best temper for trying out innovative and often risky classroom strategies. A study of science teachers' needs for professional development commissioned by the Government Office of Science and Technology and conducted by researchers from King's College London (Dillon, Osborne, Fairbrother, & Kurina, 2000) reported that teachers had strong views about the PD they received and what they felt they wanted. Amongst many such views, items relevant to the present discussion were that teachers:

- wanted more time and funding devoted to personal PD, rather than PD driven by institutional imperatives;
- were dissatisfied about how their needs are determined;
- felt the negative impact of short-term poor quality INSET.

Bolam (1982) p, 219, quoted by Hopkins (1986 p. 8), reports that

> A recent review of educational thinking and practice in OECD member countries concluded that 'Whatever their traditional approach to innovation, a number of member countries are beginning to recognize weaknesses in the centre-periphery or top-down models.

Our own position, based on all of the experiences described in earlier chapters of this book, is that while one obviously needs to respect the views and perceived needs of the teachers one is working with, a totally teacher-centred or bottom-up approach should be tempered with a perspective that sometimes headteachers and university researchers do actually know more about what is required for effective teaching for (for example) the development of higher order thinking than do the classroom teachers. One has to be careful when attempting to respond to *needs* that one is not instead responding to *wants* – a very different thing. Remember Guskey's (2000) point that an inordinate amount of time spent at the beginning of a project trying to ensure that everyone is 'on board' before the launch can end with the ship of innovation sunk in the harbour, having gone nowhere. Although Rosenholtz (1989) found in her Tennessee elementary schools that teacher learning opportunities were associated with involvement in decision making, this was not a direct effect but a correlation arising from links to a third, latent, variable.

Successful implementation of innovative teaching methods through professional development requires a combined approach, providing teachers with information, guidance, and leadership while recognising that no outsider can impose a model.

3d. Teacher turnover

The Professional Development of Teachers is just that – it is the teachers who are directly affected while the institutions in which they work are affected only indirectly, and only as long as those teachers remain in that school. The consequence of this obvious truth is that the route to school improvement through professional development will only be effective insofar as the teachers do not quickly move away to a new job. In chapter 7 we reported an investigation into thirteen schools which followed the CASE PD programme from 1994 - 1996. We discovered that in four of them the materials and methods had been almost completely lost by 1998. In one of these, all of the teachers had changed since the PD programme, and in another only one teacher in the school had ever been involved in the PD. In another four of the fourteen schools, teachers reported something of a struggle to maintain the use of the methods in the face of staff changes. Change is only as permanent as the individuals you have worked with.

The situation is complicated, however, by an interaction between professional development and teacher turnover, and here we have experienced two effects which operate in opposition to one another. A negative effect of professional development is that people who are leaders on the PD programme in a school are those who are most likely to gain promotion, often moving schools in the process. We have been struck by the number of CASE co-ordinators who have gone on to become heads of departments , advisory teachers, or Key Stage 3 consultants. It seems ironic that the PD leads to a loss for the school, but we should not lose sight of the fact that these individuals, from their more senior positions, are now more likely to promote the spread of CA methods in the wider constituencies for which they are now responsible.

We have also observed a direct positive effect of PD on teacher stability. There have been a few cases of this at the secondary level but it as the primary that it has been most obvious. In chapter 9 we have recorded the evidence of teachers who have chosen to stay in a particular school and local authority just because of their involvement with the CA PD programme. As an example, one CASE teacher from Australia chose to postpone her return for two years because of the professional satisfaction she experienced in becoming a teacher-tutor for cognitive acceleration.

Clearly the interactions between the PD, personal circumstances and personalities, and the school and LEA environment are complex and not yet well understood, and while the case studies of chapter 8 and the investigations of chapter 9 do throw light on the issue of PD and teacher turnover, more work in this area would be rewarding.

In concluding this section on school environment variables, we should note Rosenholtz (1989)'s finding from her large scale empirical study that in learning enriched schools teachers typically say they never stop learning, while in learning impoverished schools they say it just takes 2 or 3 years to learn to teach! The latter are also more likely to see teaching skill as 'natural'. A corollary is that in learning enriched schools, teachers see PD as continuous and often self-driven by experimentation and reflection, as well as from conferences and INSET, while in learning-impoverished schools PD is seen as finite, to learn a particular skill or technique, and is perceived as 'provided' by outside sources. Critically, there is a significant relationship between the school's learning opportunities and students' mathematics and readings scores: here we have a relationship established between the school's learning environment for teachers and the real outcome variable of student achievement.

A MODEL

We are finally in a position to link these variables into a model of effective professional development. This model (figure 11.1) builds on the elementary ones developed in chapter 6, recognises those proposed by Guskey (2000) and Bell & Gilbert (1996), but attempts to add both breadth and specificity from the evidence presented in this book and that drawn from the literature.

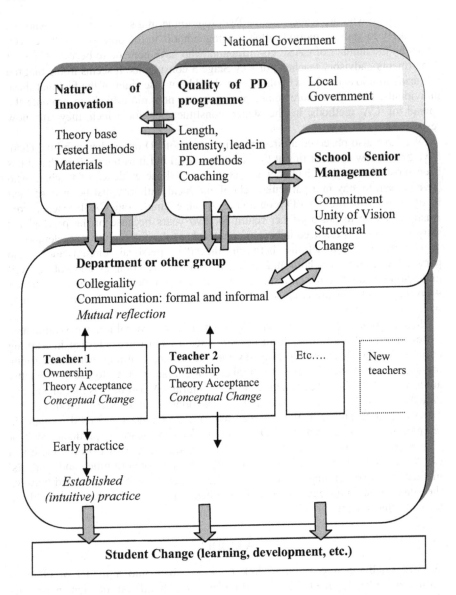

Figure 11.1: A model of factors which influence the effectiveness of professional development

Teachers are shown in a group, emphasising the importance of the community of practice, although in the end it is each teacher operating in their own classroom who brings about the changes in students. The three blocks which interact directly with the teacher group are the senior management, the quality of the professional development, and the nature of the innovation itself, summarising the positive features identified in this chapter. Deeper-level factors, those that correspond to the three-strand theoretical model outlined early in this chapter, are shown in italics. It does not make sense to draw many individual arrows of causality amongst all of the individual variables since the system is too complex, but it is useful to indicate at least the two-way interactions between each of the four main blocks. After I had presented a draft of this model at one or two conferences, comments from participants suggested that I should add also the influence of local and national governments in providing funding and creating an environment conducive to change. These appear in the background.

A critical feature of this model is that *each* of the four main blocks has to be set in a positive condition for the PD to be effective. If any one of them operates negatively, there will be little or no effect on teachers, and therefore on students. Thus although the blocks are not shown as a chain, they do have the chain characteristic that if any one of them is broken, the whole system is useless. In the same way, within the blocks *each* of the variables listed is essential to effective professional development. This chain characteristic does not apply to the background variables of government influence. Whilst support in the form of policy, planning and finance from government can have a very positive influence on the effectiveness of professional development, it is not essential.

We believe that this model "works" in the sense that it is consistent with the significant body of empirical evidence described in part 2 of this book, and with the extensive literature on professional development. We believe also that it will meet the more stringent test of a theoretical model, that it is not jut consistent with existing evidence, but that it can be used to predict PD effectiveness from an analysis of the levels of the each of the factors built into the model. In the next chapter, we must turn our attention to how teacher leaders, heads, local education authorities, professional PD providers, and national governments might use this model in a proactive way to maximise useful educational change through professional development.

12. EVIDENCE-BASED POLICY?

Policy makers sometimes bemoan the mealy-mouthed responses of academic researchers to apparently simple questions: 'Do large classes damage the quality of learning?' 'What is the effect of selection or streaming (tracking) on education?', 'How do we improve the students' deeper level comprehension of subject matter?'. To such questions, we tend to answer 'It depends ...' and then go off into a ramble about the context and what do you exactly mean by 'quality', or whatever. The problem for the policy maker – whether a headteacher or a Minister of Education - is that she or he has to take action now, in response to immediate demands of the school or of the country. 'More research needed' is not an acceptable answer and action has to be taken on the balance of probabilities, on the state of the art as we know it now. A partial and incomplete model which contains many uncertainties is better than no model at all, especially if the researcher can provide some estimate of just how uncertain each relationship in the model is. Uncertainty is no excuse for the wilful disregard of such evidence as there is, nor for the selection and misreporting of research data to meet a perceived political prejudice. There seems to have been a recent example of such misrepresentation amongst some American politicians who have attempted to curry favour by claiming that initial teacher education is virtually a waste of time, selecting as a foundation for their claim a report which breaks all of the canons of academic review (Darling-Hammond and Youngs, 2002). Less culpable, but still scary because it is actually enacted in law, is the attempt to define what counts as respectable educational research in terms of 'scientific method', as with the US Federal Legislation called 'No Child Left Behind' (Pub. L. No. 107-110). Feuer, Towne et al. (2002) and the following discussion papers in the same journal number offer a full account. The policy maker has the right to act on the basis of incomplete data or on models which are far from being fully determined, but still has a duty to make the best possible use of what research data is available.

We would claim that the model offered at the end of the last chapter encapsulates much of the knowledge that has accumulated over the years about factors which make professional development more or less effective, and that notwithstanding some uncertainties of conceptualisation and of the strength of relationships, it offers practical guidance for those responsible for managing and financing the professional development of teachers within their schools, local authorities, school districts, states, or countries.

In the first section of this chapter we will consider application of the model at school level, and then extend this to local government and national levels. We will then consider briefly some aspects of the quality and availability of professional development in the United States (since this is where the bulk of the literature on the subject originates) before critically assessing recent developments in professional development policy in England. We are not, therefore, offering a comprehensive international perspective on professional development policy, but rather an example of how our model may be used to assess the likely efficacy of a policy.

THE HEADTEACHER'S CHOICE

Headteachers have immediate responsibility for the professional development of the teachers in their school. In the UK they now have real control over their budgets and a reasonably generous allowance for continuing professional development. With few exceptions, they see PD as one of the important routes to raising and maintaining educational standards in the school, but inevitably encounter dilemmas in how to prioritise demands on the PD budget. Do you buy in a few star speakers with their smart Powerpoint presentations and engaging manner? Do you buy a weekend in a country hotel for the staff to consider their own developmental trajectories, individually and as a group? How can you balance the needs of the whole school, of departments, and of individuals? And how do you evaluate the effectiveness of the PD that you do buy?

We suggest that the model we presented in the last chapter (figure 11.1) offers some help in answering such questions. These are some of the criteria for prioritising PD spending:

- Starting in the middle of the model, the centrality of collegiality suggests that programmes which encourage groups of teachers to work together have to be more valuable than those which concentrate on individuals. A corollary of this is that the headteacher has to create the space and the atmosphere for teachers to communicate with each other, to share beliefs and experiences and to engender a wide sense of ownership.
- Looking at what a PD programme offers, ask about theoretical bases, whether the programme shows any evidence of effect on students, and whether the teachers who are to use it find the materials accessible and relevant.
- As for methods of delivery, look sceptically at any programme which claims to lead to changes in students' achievement, motivation, or other characteristics but which offers just a short one-off intensive course. Effective programmes must provide for follow-up which explores implementation and actually assists teachers in trying new methods in their own classrooms. Generally do not expect much real change from programmes which do not have a facility for contact with the providers which remains available for at least two years.
- Look to yourself: what mechanisms do you have for sharing your vision with your deputies and other senior teachers? How do you resolve differences? Are you prepared to make some structural changes to the timetable and/or to school and department development plans to maximize the chance of an innovation becoming a long-lived feature of the school? Are you prepared to take the long view?

In a nutshell, money spent on PD which offers instant success, which does not support teachers in their classrooms, whose aims have been imposed on teachers, and for which time for reflection has not been made available, is likely to be completely wasted. It would be better spent on re-decorating the reception area, getting a few new pot plants, smartening up the school brochure, and attracting parents with window-dressing rather than quality education (see Gewirtz & Ball, 1995 for an account of the school as a market).

The Local Authority

In England at least, local education authorities (LEAs) have lost much power in the devolution of funding to schools, but they do retain an important overview of all of the schools in their area, and have become the vehicle through which the present (2003) government has chosen to channel a series of 'Strategies' aimed at numeracy and literacy in primary schools, and at mathematics, English, science, and 'foundation subjects' in secondary schools. We will consider the PD implications of these Strategies in more detail later in this chapter. Apart from the Strategies, LEAs act as both providers and purchasers of professional development for their schools. We suggest that our model of effective PD may be useful to them both as a guide to designing their own PD provision and in choosing which programmes from other providers are worthy of support. LEA advisory teachers are in a particularly strong position to offer long-term continuing support to schools who are implementing change and thus have potential both to ensure the deep adoption of innovations which they propose and to support the continual refreshing of programme provided by others.

LEAs may draw similar lessons from the model as were suggested for headteachers in the last section, with respect to the choice of innovation to introduce and the quality of PD associated with it. The model may also be used to suggest something about the way in which LEAs can most effectively form partnerships with providers to offer effective PD for their schools. As described in chapter 4, we have formed educationally valuable arrangements with many local education authorities who have introduced cognitive acceleration into their schools. By combining our CA expertise with their expertise in pedagogy in general and in school-based PD, many active 'nests' of cognitive acceleration have been formed through the UK, nests from which now experienced CA tutors can venture into nearby LEAs to pollinate them also.

There are dangers also in LEA enthusiasm and the foci of these dangers, as well as of the successes, can be identified in the PD model. One relates to ownership. An enthusiastic Director of Education with some disposable funds - perhaps from one of the inner city initiatives - seeing the apparent success of cognitive acceleration in raising achievement, may simply buy in the PD for some or all of the Authority's schools. We have seen this happen and, in the absence of adequate consultation with the schools, we have seen it fail. If neither headteachers nor science departments in the schools have constructed for themselves a concept of CA as a valuable approach to PD and student achievement, they have no commitment to make it succeed. As it is given to them on a plate they do not reject it outright, but at the same time they make no real effort to take it on board since it was never their idea and they have no sense of ownership of the project. Where Authority based introduction of the innovation has been most successful, headteachers and heads of department have been consulted at length in advance, there have been awareness-raising sessions, schools have been invited to bid to join, and have also been asked to contribute financially as evidence of the seriousness of their intent. In this way both the ownership issue, and the development of headteachers' commitment (a key factor in the model) can be addressed, and local education authorities (and we suppose school districts, in spite of their very different constitution) become powerful channels, facilitators, and maintainers of effective PD for their schools.

NATIONAL GOVERNMENT

The United States

We will look briefly at some comments from the United States about the implementation of effective professional development of teachers, since so much of the evidence on which our model is based comes from there. As well as drawing from the literature, we can call on many years of attending education conferences in the USA and more extended sojourns in education departments of universities from UC Berkeley to Harvard. Let us start with a quotation from Matthew Miles:

> A good deal of what passes for 'professional development' in schools is a joke
> ...radically under-resourced, brief, not sustained, designed for 'one size fits all',
> imposed rather than owned, lacking any intellectual coherence, treated as a special add-
> on event rather than part of a natural process, and trapped in the constraints of the
> bureaucratic system... p. vii of the foreword to Guskey and Huberman (1995).

This seems to be a harsh judgement, and yet it is reinforced by others. Elmore (1996), quoted by Earl et al. (2003) argues that even the most successful efforts to change educational practice have rarely influenced more than 25% of classrooms in the U.S. Rosenholtz (1989) concludes her study of Tennessee schools with the claim that increasing national and state control over teachers, as a response to perceived incompetence ("A Nation at Risk", National Commission on Excellence in Education, 1983), pays lip service to increasing professionalism but takes action to reduce it.

> Many of the recently passed reforms try to regulate both the content and process of
> education in the hopes that teacher proof instruction will increase the quality of
> schooling. Legislators and administrators seek to enforce hierarchical control over
> teachers through such routine devices as management-by-objectives, standardized
> curriculum packages, and minimum competency testing. p. 214

Thus it becomes difficult to attract able teachers. She quotes (p. 26) small rural districts confronting severe teacher shortages which sometimes desperately, illegally, and therefore surreptitiously, recruit teachers from institutions which grant two year Associate Teacher Certificates.

Our own experience of trying to introduce CASE into one American school district was quite instructive (Forsman, Adey et al., 1993). The science inspector was enthusiastic, he identified a teacher who took on the project whole-heartedly, internalising the model and developing many new activities within it, to form a complete high school freshman science programme. I was invited to run inservice courses for science teachers from the nine schools involved but was a little surprised at the apparent world-weariness of many of the teachers, who gave a distinct impression of having seen one innovation after another with no lasting effect. This was a distinct contrast to the reception accorded our programme in other parts of the world. I was informed that attending inservice courses was a necessary condition for stepping up the pay scale. We looked at pre-test – post-test gains in a pre-CASE year and in the first CASE year. The latter were significantly higher than the former in all nine schools. This seemed encouraging, but when we looked in more detail we found that there had been no change in post-test scores from the non-CASE to the CASE year. What had changed was the pre-test scores, significantly depressed in the second year. I asked about demographic changes which might account for this. There were none, but my science inspector friend explained: "These teachers are very accustomed to high-stakes testing. If they think their appraisal depends on getting better gain scores, they know well enough how to depress the pre-test scores".

There is no doubt that there are excellent examples of professional development in action in the United States which counter these rather gloomy snapshots. Examples include Success for All (Slavin, 1990), the well known New York City District 2 project (Cooper and Boyd, 1999), the Community of Learners work of Magdalene Lampert, Anne Brown, and others (Bruer, 1993), work on science teacher leaders in North Carolina (Nesbit, Wallace et al., 2001) and Virginia Bill's primary mathematics work in Pittsburgh (Resnick, 1992). More recently we have also read in Earl, Watson et al. (2003) of successful State-wide professional development in Connecticut (Wilson, Darling-Hammond et al. 2001). But if the overall picture is one of malaise, or perhaps more accurately one of many small successes oddly isolated form one another, can some general cause be identified? Dale Mann (1988) has suggested that one reason might be constitutional reverence which has always been paid to local decision-making. "Financial support for public schools in the USA has always been a vehemently local responsibility" (p. 5). Notwithstanding the $250 million set-aside for effective schools provided under Ronald Reagan following the "A Nation at Risk" report, the policy legislation treats the educational process as a black box, offering money and demanding better teacher qualification to achieve higher outcomes, while remaining agnostic as to how this might be achieved. That a higher proportion of teachers belong to labour unions than any other occupation – and there are 2 million teachers – means that "It is not surprising that elected officials have checked their zeal for educational reform at the classroom door" (p.6). In a system of 18000 school districts each of which is determined to demonstrate its independence from State - let alone Federal -

interference, it is difficult to image how professional development for school improvement could be anything but patchy.

The role of the universities in this process might also be worth some scrutiny. In the United Kingdom a common career path for a professor of education will have been from first degree, through a substantial period as a schoolteacher, then to teacher education and a PhD on school-based research. In the US, professors of education are generally less likely to have had significant school teaching experience themselves, and much of the school practice of student teachers is tutored by graduate students who have little practical teaching experience themselves. US professors of education may, on average, have better academic credentials than their UK counterparts, especially with regard to research methods, but they are further removed from school teachers in terms of both experience and status. I realise that this analysis may cause offence to friends on both sides of the Atlantic but if there is any truth in it, it puts American education researchers at something of a disadvantage in trying to bring about change in schools, since from the teachers' perspective the researchers' motivation may be perceived as not so much the improvement of education for its own sake, but rather the completion of a research study which will yield academic papers, a book, and maybe further research grants. I am not saying that this characterisation is accurate or fair, but that it is the perception of many American teachers whose status, as Fullan and Stiegelbauer (1991) describe so graphically in their chapter 7, is remarkably depressed.

In relation to our model of effective professional development, there may be no problem with the nature of the innovations being generated in US universities and States education authorities, the level of senior management commitment may be little different from the UK, and on average school-level factors are likely to be similar. There does, however seem to be a problem with the quality of PD programmes - in particular, the requirement for long-lived programmes which include coaching. These have been called for for years by Bruce Joyce, Michael Fullan, Matthew Miles, and many others, but it often seems to be short-circuited for reasons of finance and political and academic short-termism.

England

There is an irony about British politics and education. Conservatives are normally seen as philosophically predisposed to laissez-faire economics, local control of services, and the dismantling of central government power, while Socialists are characterised as bent on central planning of the economy and services. But in education from the 1960s to 1997, precisely the opposite policies were followed by successive British governments. Labour tended to leave educational decisions to local education authorities, there was no national curriculum, end of school examinations were set by independent boards with no government control, and schools were constrained only by their perceptions of the needs of their students for university entrance or vocational training. Margaret Thatcher's Conservative government changed all of that, with the introduction of a national curriculum, examination boards given mandated 'standards' to adhere to, and local authorities emasculated by having their funds withdrawn and passed directly to schools. Graham, Gough, & Beardsworth (2000) offer a critique of this attempt to control the provision of professional development (pp 4–8). At the time there was some opposition from Labour spokespersons who raised fears that such measures could undermine teachers' professionalism. Since the election of the Blair Labour Government in 1997, however, this irony has been ironed out. The central control of education built up through the Thatcher years has been accepted wholeheartedly and in many ways made even firmer. Brown, Millett, Bibby, & Johnson (2000) chart in detail how, partly by the mechanism of retaining the same Chief OfSTED Inspector, the incoming Labour Government of 1997 built on and strengthened the centralist and traditionalist educational policies of their predecessors.

In this section we will look at the influence on the professional development of teachers, not of centralisation as such, but of the policies which have been centrally determined and which centralisation has allowed to be widely implemented. We will focus on the situation in England specifically, since the arrangements in Wales are somewhat different and Scotland and Northern Ireland have quite different educational systems. The purpose is to illustrate the application of our model of professional development to national effects, rather than to try to catalogue a range of such instances. The emphasis will be on the professional development provision and methods within the "Strategies", the mechanism by which the government aims to improve educational standards, although we will need to look first at the policy which has been developed on continuing professional development.

Continuing Professional Development Policy
First of all, one must commend the Blair government, with David Blunkett as its Secretary of State for Education, for even realising that a policy on professional development was worth having. Soon after the publication of a consultation document on the subject in February 2000, Bob Moon wrote in the *Times Educational Supplement* (March 17th, p. 17):

> It is nearly 30 years since the last national debate on professional development. The fabled James committee set out ambitious proposals for a national framework of in-service education. Margaret Thatcher, as Education Secretary, invited the committee to the Dorchester, thanked them for their efforts, and the proposals were quietly dropped.

The policy document which emerged following the consultation (Department for Education and Employment, 2001) necessarily addresses the broad sweep of issues concerned with effective professional development and cannot be concerned with the introduction of particular innovations into a school or school system. It aims to provide a national framework within which PD may be supported and evaluated. Nevertheless, there are some specific recommendations and actions (with funding) proposed, and we will consider each of the six main topics in the light of our model for effective professional development.

1 *Increased funding, for example for teachers' research scholarships, 'professional bursaries', and sabbaticals*. These are aimed predominantly at the development of individual teachers and so will only have a place in our model if they are used as part of a school-wide or possibly Authority-wide programme. There can be many useful outcomes of individual professional development, but significant change in schools is not one of them.

2 *Select development activities which have impact*: this is a rather catch-all section, but it does recognise, verbally at least, the value of coaching, of theory-driven change, of reflection, and of learning from others. We must applaud all of this, and wait for the mechanisms by which they will be implemented. One of these is promised to be though the 'Strategies' to be discussed below.

3 *Improving provision*: The tactics here include specifying a code of practice for PD providers, developing and supporting Advanced Skills Teachers who have a PD responsibility in their own and in neighbouring schools, and funding award-bearing PD courses. All of these address the 'Quality of PD programme' element in our effective PD model and so create the framework in which high quality PD programmes can be developed. The associated Code of Practice for Providers (Department for Education and Employment, 2001) is unexceptionable, but focuses more on management issues (planning, venue, monitoring, clarity) than on the nature of high quality PD. It may be argued that this stance recognises the professionalism of providers rather than trying to specify one or a limited number of 'best methods'.

4 *Identify and spread good practice*: the concern here is that there are a lot of good things going on, but that there is not enough dissemination of these good practices. Inevitably a website is proposed[1], but there may be a problem of ownership here. Neither our experience nor our perusals of the literature reveal the borrowing of professional development practice by one school from another to be a widespread or particularly useful practice. It has been a constant theme throughout this book that change in teaching practice is slow and uncertain, requiring much collegial and/or outside support. It is thus unlikely that any account downloaded from the web will have real impact on pedagogy. The idea of some schools becoming 'professional beacon schools', offering support to other schools in their region, does seem to have more mileage in it. It is an idea that (Fullan, 1995) has proposed and seems well worth trying, provided that the inherent competitiveness between schools (strongly fostered by the Conservative government with its emphasis on league tables of exam results) can be overcome.

5 *Raising expectations*: this section unashamedly proposes exerting pressure on teachers through performance management and OfSTED inspections to take responsibility for their own professional development. Our experience in the United States of working with teachers who attend PD programmes because they are

---

1 A hour spent searching the site in March 2003 took me around in many circles. At one point I did find some case studies, but none were dated later than October 2002. On my next visit I could no longer locate these, but I accept that this may be my problem, not the site's. Access is said to be through http://www.teachernet.gov.uk/.

required to do so in order to move up the pay scale does not encourage us to believe that such pressure is likely to be productive, except insofar as it points teachers in the direction of reflecting on their practice and trying new pedagogies. We have previously quoted Guskey & Huberman's (1995) finding that commitment often follows change in practice, but this is a tool which must be used with extreme caution.

6 *Research into effective PD*: Naturally we applaud the intention to fund research into effective professional development.

In concluding this review of the government's policy towards CPD, it should be asked: are there any elements of our model of effective PD which have not been addressed by the policy? The most obvious omission is any reference to the nature of the innovation, or development, to be introduced by the professional development. The policy focuses on delivery and dissemination, but says little about just what is to be delivered and disseminated. We have frequently referred with approbation to Fullan and Stiegelbauer's (1991) insistence that we must seek evidence that an innovation being introduced is, in fact, worthwhile. Although this may seem obvious, it is certainly the case that much PD delivered in schools in the UK (and, we suppose in the US also) is actually designed to develop pedagogic practices for which there is no evidence of effect on students. It does not seem much to ask that a government policy on professional development, and its associated code of practice for PD providers, should spell out a requirement that the proposed pedagogical development is actually useful.

With the exception of this odd lacuna, and the minor quibbles noted in our review, the CPD policy has to be seen as a welcome document which opens up many possibilities for the further development of high quality professional development in England. Centralisation does have its advantages.

The GTC Teachers' Learning Framework

An interesting counterpart to the government's policy on PD is provided by the General Teaching Council for England (GTC), established in 1998 to reflect to the government the professional views of teachers and to guarantee and maintain professional standards. *The Teachers' Professional Learning Framework* (General Teaching Council for England, 2003) is a blessedly slim document the meat of which is about 55 bullet points summarising teachers' entitlement and mechanisms for effective professional development. The emphasis throughout is on collegiality and in this and in other respects it accords very closely to the principles encapsulated in the central 'department or other group' box in our model. It is offered as "...a tool, for the school and individual teacher, to help plan professional learning" (p.2) and in our judgement is likely to be a powerful tool for this purpose.

The Strategies

It sometimes seems that the profile of education in politics in England has been on an inexorable rise ever since the Butler education act of 1944, which provided for universal secondary education in Britain. Tony Blair's declaration in 1996 that if he was elected, his government's priorities would be "education, education, and education" must have been judged to play to the voters' perception that, notwithstanding the years over which a national curriculum had been instituted and revised, and regular national testing had been established at ages 7, 11, and 14 in addition to the older tests at 16, and 18, the system was still failing to deliver adequate standards. We have mentioned previously Hopkins & Lagerweij's (1996) four decades of the search for raised educational standards: the 1960s when printed curriculum materials were expected to do the job; in the 1970s there was failure and hand-wringing; the 1980s saw studies of school effectiveness and some success in identifying key variables, and the 1990s became the era of managing change, the school improvement movement. The Strategies promised to bring the power of school improvement to bear at a time when political demand was at an unprecedented level. The Blair government with David Blunkett in charge of education picked up on work that had been started by their Conservative predecessors on literacy and numeracy standards in primary schools, and turned them into the National Literacy Strategy (NLS) and the National Numeracy Strategy (NNS) which went nationwide in 1998 and 1999 respectively. These were followed by a set of Strategies for Key Stage 3 (12 – 14 year olds): English and mathematics rolled out in September 2001, with ICT, foundation subjects, and science going national in 2002, in each case after one year of piloting in a limited number of schools. An immediate concern about this programme was the rapidity with which the Strategies, which were intended to have radical effects on methods of teaching, were implemented. They were written by small teams and then imposed on the nation's schools with only one year pilots and no time for real formative evaluation and revision. The government was in a hurry and was not to be deflected by academic niceties.

Described by Earl et al. (2003 p.11) as "the most ambitious large-scale educational reform initiative in the world", the Strategies are aimed unequivocally at raising academic standards, interpreted as levels achieved in national tests. They go far further towards defining both content and teaching methods than do either the programmes of study of the national curriculum, or the pressure of assessment of the Key Stage tests. The Strategies provide 'Frameworks' which, alongside detailed schemes of work published earlier, provide details of the curriculum content to be covered over every one of the first nine years of statutory schooling. Moreover, they offer detailed guidance to the teaching methods to be employed which it is claimed will cohere a disparate range of established good teaching practices. Although schools are not required by law to follow the Strategies (unlike the national curriculum and Key Stage tests) the expectation of OfSTED inspectors is such that only the most confident schools could afford to ignore them. In effect, all of the detail and more of the first version of the national curriculum, which went through a series of revisions to reduce detail in response to schools' fury at having such detail legislated, has been returned, albeit with less than statutory force.

Within the subject matter of this book, it will be proper to focus on the professional development opportunities and provisions of the Strategies, rather than on the particular subject matter content and teaching methods which form the bulk of their published manifestations. (And 'bulk' is the right word: each of the two primary and five secondary Strategies is represented by boxes full of booklets, multi-coloured leaflets, large folders of objectives and of training sessions and videos) However, since an important element in our model of effective PD is the quality of the innovation, the pedagogical methods spelled out by the Strategies demand some attention. A universal shape of lessons is proposed: quick starter activity, main teaching activity, and a final 'plenary', that is, whole-class round-up. The justification for this particular shape and for much of the detail of the methods proposed is said to rest on evidence from OfSTED inspectors and from 'research' although – perhaps in the interest of teacher friendliness – references to the research are limited. In one area in which I would feel competent to comment, the science Strategy, I would have some reservations about the directions given on, for example, misconceptions and modelling. The research evidence is eloquent in its failure to report success in shifting children's misconceptions through any form of regular teaching process (e.g. Driver, Squires, Rushworth, & Wood-Robinson, 1994; Pfundt & Duit, 1988; Vosniadou, Ioannides, Dimitrakopolou, & Papademetriou, 2001). In an extensive review, Brown, Askew, Baker, Denvir, & Millett, (1998) have thrown serious doubt on the claims that the pedagogies espoused by the National Numeracy Strategy (and by extension the Strategies in general) have any real evidential foundation. Only in the foundation subjects does there seem to any readiness to acknowledge and build on the research evidence - but then I would say that, wouldn't I, since they draw on our work on the teaching of thinking and on my colleagues' work on formative assessment. The point here is simply to question the validity of the evidence on which the teaching methods are based. Looking at the Strategies through the lens of our model of effective professional development, within the 'Nature of Innovation' window there appears to be inconsistent use of theory bases and serious doubts about the extent to which the methods can said to have been tested, in the normal research use of that term. On the positive side, the materials are comprehensive, well produced, and easy to use.

The Strategies come with a heavy commitment to funding, much of it for professional development. In contrast to the content and teaching methods proposed, management of the PD programme is treated rather flexibly and so it is cause for celebration that a comprehensive and expensive system of PD does form an integral part of the Strategies. On the other hand the PD programme does follow a classical cascade model. There are comprehensive Training[1] packs which specify in detail scripts which are to be followed, with OHTs, Videos, and CDs containing PowerPoint presentations. These have been written by small teams and are then

---

[1] My colleague Anne Robertson is adamant that Training is for dogs, and she refuses the use the word in connection with teacher development. I simply copy the language of the Strategies.

implemented, for each Strategy, through a national director and four or more regional directors to consultants in each LEA, who deliver the sessions to one lead teacher in each subject from each school, who finally passes the baton on to colleagues in school. As is well known, this approach has its problems. An OfSTED evaluation (Office for Standards in Education, 2003) after the second year of the national use of the English and Mathematics Strategies at KS3 reported:

> The challenge thereafter was for lead teachers in each strand to disseminate the ideas and approaches to colleagues who had not attended the training. Generally this was good in English, but in a third of the schools mathematics teachers were not sufficiently well informed about the Strategy. Dissemination in schools was unsystematic in science, and often weak in foundation subject departments with large numbers of teachers. Dissemination to specialist ICT teachers was effective but did not extend so well to non-specialist teachers of ICT. (p. 3)

For the primary school Strategies, the cascade connects at school level with numeracy and literacy co-ordinators who are responsible for dissemination across all Years but also with 'Lead Teachers' who are similar to the teacher-tutors we developed in Hammersmith. They continue to teach in their schools but also have some time available to help in other schools in their neighbourhood. In the majority of cases, where management is strong, this seems to have been more successful than the secondary school experience.

At the school level, the approach is based on schools completing an 'audit' of their needs, equivalent to the needs analysis phase of introducing an innovation. Selection of sessions from the training packs is to be made in the light of the audit. This in turn means that the activities in the training materials do not show a sense of progression from one session to the next, and no opportunity seems to be provided for reflection on difficulties and successes encountered so far in the implementation process.

The consultants both respond to requests from schools for help in implementing the Strategies, and are proactive in approaching a selection of schools in their LEA to offer 'additional support'. This selection is based on a mixture of schools which seem especially to need help and those which provide evidence that they will make good use of any support offered. Headteachers, heads of department, 'Strategy managers' and others attend half day and one day briefings and INSET sessions run by consultants or by other PD providers. Thus the programme, while centred on the training pack designed to develop the pedagogies specified in the frameworks, is supported by a variety of mechanisms for drawing in expert advice on the effective use of the training sessions.

From the start of the introduction of the Strategies, there has been some concern about the rigidity with which the implementation appeared to be directed. There is no doubt that some senior officers in the Department for Education and Skills, not necessarily professional educators themselves, took a strong line against any flexibility being encouraged in the implementation, or allowing any 'interpretation' of the training programme as laid down in the printed and video material. The extensive documentation is characterised by a "prescriptive and certain voice" (Brown et al., 2000 p. 462). For example:

Training has been least successful when rambling presenters have underused the script, video, OHTs and handouts and not made the key messages explicit (Department for Education and Skills, 2002 p. 12).

On the other hand, Earl et al. (2003) talk of a persistent misconception about the inflexibility with which the Strategy methods are to be implemented, and report that 95% of NNS and NLS consultants feel that they do have flexibility in implementing training. I have explored these claims with perhaps half a dozen consultants and two regional directors in the science and foundation subject Strategies, and every one has reported that the old rigidity is in fact now rare, and only seems to occur when a consultant (or even a regional director) is unsure of themselves and lacks the confidence (one might say the professionalism) to make the materials their own and to deliver the central message in their own way. Notwithstanding the dangers inherent in allowing teachers to make their own sense of an implementation (Huberman & Miles, 1984), this is an issue on which the whole success of the Strategies could hang. It has been emphasised throughout this book that all of the evidence we have accumulated and all that we have reviewed points to the centrality of professionalism in professional development. The three keys of teachers' attitudes and beliefs, their sense of ownership, and their intuitive practice are never developed by sets of procedures and instructions to be implemented, however detailed and well-meaning the procedures may be. Insofar as there are remnants of rigidity in the delivery of training within the Strategies, they will fail to meet the criteria of effective PD methods shown in the 'Quality of PD' window of our model and thus will undermine Strategy implementation. Fortunately Brown et al. (2000) point to a number of inconsistencies in the Strategies' various prescriptions which make rigid adherence logically impossible, and in any case they concur with my informants' optimism in believing that teachers will always interpret what they are asked to do:

For the policy writers of the Strategy, losing control of meanings of their texts may be difficult to accept. Prescription of the content of the training materials at numeracy consultant as well as at school level has sought to minimize the possibility of this happening. The provision of videos, script, and Strategy-approved artefacts might be interpreted as controlling what teachers see, hear, and use. But teachers' own experiences, values and purposes will all play a part as they process what they see, hear and are offered, and make sense of it in their own ways. P. 469

There are two more items in the 'Quality of PD window' of our model: length intensity and lead-in, and coaching. In the Strategies, the former is largely in the hands of the schools. Funding will be provided over two or three years and if the key personnel – consultants, headteachers, heads of department, Strategy managers, and others see implementation as an ongoing process over this period, there is a good chance that this requirement for effective PD will be met. However, the value of a long-lived implementation programme will only be reaped if teachers perceive it as allowing them gradually to develop their practice with reflection on successes and 'dips', rather than as a series of individual ideas introduced one after another. As for coaching, there are many references to the benefits of peer coaching which run through the Strategy documents, so the message from Bruce Joyce and his colleagues, reiterated to the point of tedium in this book, has been taken seriously.

Since the Strategies are essentially an outside-in programme from the government to the schools, it is in our model windows of Nature of Innovation and Quality of PD that we can most easily make judgements, but the Senior Management and department windows also deserve consideration. The implementation programme does include briefing days for headteachers, and as noted above the pressure of inspection expectations virtually ensures that all must take the Strategies seriously. But does this constitute 'Senior Management commitment'? The comprehensive nature of frameworks and training materials, with their objectives, may increase the chance of shared goals between headteacher and other senior staff, and may also help to establish the structural change we see as essential for effective innovation, but the *commitment* must come from within the individual headteachers, it cannot be imposed from outside. The government's task here is one of persuasion, over and above the provision of money for training and packs of materials. The high status deservedly given within the Department for Education and Skills to Michael Fullan and his work on leadership (Fullan, 2001) signals that they take this issue seriously.

At the department level, the KS3 English Strategy at least does emphasise the need for collegiality in PD, and the requirement that a critical mass of staff need to be involved in working together to achieve effective implementation. Whether or not the cascade methods of the Strategies will achieve effective critical masses in schools will depend importantly on whether sufficient time is provided within the school for PD sessions, peer coaching, and reflection. This in turn takes us back to headteacher commitment, signalled above as a critical controlling factor.

The remaining elements in our model are teacher ownership, theory acceptance, and conceptual change, leading to the establishment of changed intuitive practice. The pessimistic view here is that the largely theory-free nature of the Strategies and the mechanistic way in which at least some DfES officers would like to see them implemented will do little to promote these important elements in teacher professional development. The optimistic view is that schools which combine senior management commitment with skilled professional interpretation of the Strategy training sessions will indeed be able to engender the deep-seated changes in teachers' beliefs necessary to underpin permanent change in practice.

To summarise this view of the Strategies in relation to our model of effective professional development, we are concerned about the absence of theory and the lack of research justification for the methods being introduced, but impressed by the quality of the materials. The cascade approach may on the face of it be worrying, but it seems that it is only the final link, from lead teacher to colleagues in school, which has not so far been made to work reliably – but it is still early days. A question hangs over headteacher commitment, on which so much depends, but the attention to collegiality and coaching is to be applauded.

What evidence is there that our concerns are justified? After all, they are based on relating the PD elements in the Strategies to a model which has only just been constructed and offered for critical appraisal. Here we can turn to a considerable amount of evaluation evidence which has been provided both by OfSTED, and by an independent team from the Ontario Institute for Studies in Education (OISE). OfSTED reports need to be approached with a different mind-set from that which an academic is accustomed to bring to evaluate a research paper. They have a pyramidical structure in which the published material represents only the tip of a mass of evidence collected in a very structured manner and in accordance with tight criteria, often cross-checked by pairs of observers. The language is exact, with words such as 'good', 'satisfactory' and 'unsatisfactory' having precise meanings which can be defended, if needs be, by reference back down the pyramid, eventually to the inspectors' immediate observation notes. OfSTED reports are written in a dense style, with a great deal of information provided in relatively few pages so that it is virtually impossible to summarise further without losing essentials of the picture, and to quote passages is to risk the accusation of unrepresentative selection. But here we will try.

The OfSTED evaluation of the second year of the Key Stage 3 Strategies (Office for Standards in Education, 2003) is based on Inspectors' visits to 126 schools, some of which were revisited near the end of the 2001-02 school year, and information culled from regular OfSTED reports, some especially enhanced to evaluate the Strategies. There is a wide mixture of successes and concerns. Virtually every one of the 156 numbered paragraphs and 27 bullet points of main findings is of the form "x is going well but there is a concern about y". Germane to this book, there is praise for the work of the consultants, but concern about dissemination within schools (see the quotation on p. 186 above). There is unalloyed corroboration of the centrality of departmental leadership and collegiality which form such important elements in our model, and of the role of senior managers:

> Most English departments made considerable efforts to review and change previous practice, where necessary, and adopt aspects of the Strategy. Key factors were the attitude and quality of the head of English department. In schools where progress was unsatisfactory, departmental leadership was often poor, or the attitude of the head of English was negative. (para. 17)

> The management of literacy across the curriculum was most successful in schools where a group of teachers co-ordinated the work .... (para. 20)

In mathematics:

> The Strategy was most effective where there was a core of capable teachers who were enthusiastic and committed to raising attainment ... (and) ...teachers had sufficient time to discuss teaching and learning and to plan lessons together. (para. 56)

In the foundation subjects

> Where the management was most effective, senior staff had a good grasp of what was happening and often took part in meetings and in the planning. (para. 130)

> In a third of the schools, the management of the strand was ineffective because of poor communication between members of the foundation team. There was limited

involvement of senior staff, and there were no plans for disseminating good practice to others in the school. (para 133).

Many more examples of the same story could be quoted from each of the subject areas. Overall, this early OfSTED evaluation of the impact of the KS3 Strategies is cautiously optimistic, but offers many suggestions for areas which need attention if the Strategies are to influence the quality of teaching and learning in the majority of classrooms in the country. As far as effects on learning are concerned, it is claimed that "There are promising signs of the effect of the Strategy on attainment" (p.3), but data included in an annex shows no measurable effect on English or Mathematics KS3 scores in pilot schools compared with controls. It is, however, pointed out that no gains should be expected until 2003 when the first Year 7 students who experienced Strategy methods come through to the test.

With the primary National Numeracy and Literacy Strategies, there should have been more time for any effects to become apparent. The evaluations of NLS (Office for Standards in Education, 2002a) and NNS (Office for Standards in Education, 2002b) are each based on surveys of 300 representative primary schools visited at least five times each, regular inspections, a telephone survey of 50 headteachers, and a special testing programme. The evaluation commissioned from the Ontario Institute for Studies in Education (Earl et al., 2003) is in a form more familiar to academics than are the OfSTED reports. Data collection methods and sampling are fully described, and the implementation of the Strategies is set in the context of the international literature on school improvement and professional development.

Given the government's focus on test results as the prime indicator of success, it is not unreasonable to look first for any evidence of effect of the Strategies on students' achievement of higher levels in the national curriculum tests, even if from a broader perspective such a measure of success may be considered rather limited.

As targets become more and more difficult to reach, they detract from, rather than support, teaching. (Earl et al., 2003, p. 48).

Very curiously, none of the three evaluation reports really comes clean on this issue. The OfSTED reports note that percentages of students reaching level 4 at Key Stage 2 rose somewhat in the year after introduction of the Strategy, but seem to have reached a plateau, and the OISE report notes that the same measure for science, which has no strategy, actually rises more steeply that those for English or mathematics. (It was my colleague Margaret Brown who first noted this comparison with science levels. Details are to be published in Brown, Askew, Millett, & Rhodes, 2003). If when one plots the data since the time that the tests began in 1996, and marks in the start of the National Literacy and National Numeracy Strategies (figure 12.1), the case for any effect of the Strategies on student achievement seems, at best, opaque. NLS and to a lesser extent NNS came it at a time when the success rates was already climbing steeply, probably due to the increasing test-wisdom of schools but maybe also because of changes in the nature of the tests. Neither Strategy has any discernible effect on achievement, as measured by the percent gaining level 4. In fact compared with science it could be argued that the NLS and NNS have actually been detrimental to students' achievement. The data is consistent with the rather conservative and behaviourist explanation that growth in scores can be attributed to testing alone. The suggestion in the OISE report that science may be affected by improved literacy is difficult to square with their finding that only 64% of numeracy, and 51% of literacy consultants agree that teachers use Strategy teaching approaches in other subjects (Earl at al. 2003 p.73) Many headteachers and teachers interviewed talked of teaching to the test, or 'hoop-jumping'.

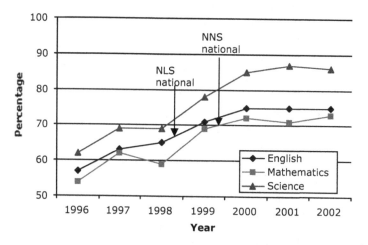

Figure 12.1: Percentage achieving level 4 or above at KS2 tests

Measurable effects on students is the bottom line of our PD model and was the purpose of the Strategies. If there is no effect, then discussing the effectiveness of the PD programme associated with the Strategies may be entirely academic - but then this is an academic book. Even if the pedagogy promoted by NNS and NLS has not yet been shown to improve learning, then it remains possible that the PD programme is excellent in its own terms, but is being used to develop ineffective teaching methods. So, we will look briefly at what the evaluation reports say about the effect of the Strategies on teaching. It is important to remember that in the following discussion, what counts as 'good teaching' is that which is defined as such by the Strategies.

The OfSTED reports say that the quality of the teaching showed greater increases in the early years of the Strategies, although the teaching of writing and of mathematics continue to improve. The NLS report describes a sea-change in the teaching of reading which NLS brought in and which is now well established. In both NNS and NLS, there is concern about the quality of the 'plenary' sessions.

(This mirrors the observation we have made over the years in CASE that the most difficult of the pillars for teachers to take on board is metacognition, reflection back on what has been learned and how it was learned.) In NLS there is a confident air about what is needed to raise standards further (for example, by increasing the amount and quality of phonics) but, in line with questions raised above about the quality of the innovation, the justification for this certainty is unclear.

As with the KS3 report (and unsurprisingly) there is a strong emphasis on the role of the headteacher. The OISE report commends a new focus on leadership capacity building, and the OfSTED reports claim that NLS is threatened in about one in ten schools (NNS one in eight) where the headteacher offers weak leadership. For the great majority, effective headteachers support the Strategies by providing strong cross-year and cross-curricular co-ordination.

The consultants' work is met with wide approval, although in-school coaching is considered more valuable than out of school INSET days:

> It was clear, however, from evaluation in the most successful schools that training which took place in classrooms through lesson observation and feedback, demonstrations, and team teaching was more effective overall than which took place away from school or during out-of-school hours (NLS evaluation para. 144)

OISE claim that the PD of the Strategies has developed from a cascade + distance learning model to one based more on coaching.

> NNL and NLS training sessions ... increasingly incorporate tasks that connect to classroom practice, often with provision for follow-up sessions in which participants can review and extend their learning. p. 41

At the same time, they are concerned that the PD doesn't provide opportunities for more than superficial learning of practice, without understanding. They report that most teachers – and there are nearly 200,000 teachers in the country - had been involved in at least one PD opportunity but the majority of these 'opportunities' consisted of using a videotape in their school.

> But, given the scale of the enterprise, it is not surprising that few teachers have experienced sustained and job-embedded learning, This, however, is the kind of learning necessary for large numbers of teachers to become competent and confident about new teaching approaches and content that may be fundamentally different from past experience: p. 91

OfSTED points to the value of co-ordinators from different schools meeting together to share concerns, often with the consultant. As with our work in Hammersmith, at the primary level it seems that cross-school meetings provide powerful collegial support which, in secondary schools, may be found within a department.

The NLS evaluation (but not the NNS) makes something of a play for the revision of the Strategy materials, in particular the Framework. They seem to put faith in raising the standards off the plateau more on revision of the Framework than on developing the PD support for teachers. On the one hand, we might read this as suggesting an attitude to educational development which is reminiscent of Hopkins

& Lagerweij's (1996) phase 1: the 1960s when curriculum materials were expected to do the job. On the other, we could applaud the implied challenge to the Strategy materials as the established best approach. Our final quotation gives grounds for optimism:

> ... a great deal has been achieved, but further progress will depend on an open critical approach to the Strategy at a national level. (para. 153 NLS evaluation.)

Earl et al. (2003) talk of the dilemma that regional directors and consultants find themselves in since they know that over-centralised direction is a bad thing, but they can't see how they could have got this far without it. They want to be more flexible, but that requires a deep sense of ownership, and that in turn requires a belief in the soundness of the methods being promoted. In the absence of evidence for any effect, such belief may be difficult to foster.

Finally, the OISE report points up a strange paradox: consultants claim that many teachers have not yet changed their practice effectively, and yet they believe that pupils are learning more, while teachers say they have changed radically, but that pupil learning is only slowly improving. Looking back to chapter 10 on methods of evaluating professional development, this paradox does highlight a difficulty with evaluation methods that rely overly on collecting opinions of participants, while by-passing the quantitative evidence that is available – which in this case suggests that no greater learning is occurring which can be attributed to the Strategies.

The worst-case interpretation of the Strategies so far seems to be that a vast amount of money has been spent on a PD programme of moderately good quality which is being used to introduce methods which have no theoretical foundation and show no evidence of having any effect on learning. Against this must be set the vast scale of the enterprise, the real vision shown by a government in even attempting to change teaching practice in such a proactive manner – not just relying on a prescribed curriculum but developing a comprehensive programme of professional development – and the fact that it is still early days. In parallel with implementation the government has not fought shy of running a proper evaluation, and the evidence at the time of writing is that the problems raised by these evaluations are being addressed. That sure beats the record of any other government initiative I have ever encountered – in the Caribbean, Indonesia, Singapore, the United States, or previous British governments, all of which introduce major curriculum innovations with much fanfare but then avoid real 'warts and all' evaluation and allow them to fossilise. Normally, it is the show rather than the substance that interests politicians.

CONCLUSION

In this final section we will review the development and application of the model of effective professional development which was presented at the end of chapter 11, and consider how it might be used as a powerful tool in the design and evaluation of any professional development programme, from the introduction of a specific teaching technique to a PD programme which is part of a national school improvement strategy.

In Part 2 (chapters 5 – 9) we presented a varied body of evidence from our own work over 25 years in cognitive acceleration and other PD-rich projects highlighting the impact of a range of factors on effective professional development. Many such factors were identified, and in chapter 11 we tested each against the findings of others and provided evidence for them being commonly observed and not idiosyncratic to our own experience. The factors are grouped here under four headings in a formulation slightly different from, but equivalent to, that presented in chapter 11. It has been established that each of the factors is necessary to effective professional development:

1   The Innovation
    1a   has an adequate theory-base
    1b   introduces methods for which there is evidence of effectiveness
    1c   is supported with appropriate high quality materials
2   The PD programme
    2a   is of sufficient length and intensity
    2b   uses methods which reflect the teaching methods being introduced
    2c   includes provision for in-school coaching
3   Senior management in the school(s)
    3a   are committed to the innovation
    3b   share their vision with the implementing department leaders
    3c   institute necessary structural change to ensure maintenance
4   The teachers
    4a   work in a group to share experiences
    4b   communicate effectively amongst themselves about the innovation
    4c   are given an opportunity to develop a sense of ownership of the innovation
    4d   are supported in questioning their beliefs about teaching and learning
    4e   have plenty of opportunity for practice and reflection.

Each of these factors, built into the model of figure 11.1, forms an essential link in the chain from the intention of the PD provider to changes in students. If one of the links is weak or broken, there is little or no opportunity for providing compensation by strengthening a different link. Looked at this way, the process of effective professional development is both complex and fragile and it becomes surprising not so much that it fails so often, but that it is occasionally successful.

The model is offered as a tool for those developing professional development programmes, who can use it to interrogate their plans for the provision they make for each of the essential factors. But we have shown in this chapter 12 that it can also be used as an analytical tool to investigate the elements in an established innovation and identify potential sources of weakness or causes of ineffectiveness. We have used this analytical lens extensively on England's 'Strategies' not only because they are "the most ambitious large-scale educational reform initiative in the world", or because they are of immediate importance to the education of millions of children, but because they offer a complex and current example to demonstrate the power of the PD model as an analytical tool. It will be clear that the model does not impart a rose tint to the systems it interrogates, but that it is able to pinpoint specific areas within the system which would benefit from closer attention as the implementers continue to seek for improvement.

Finally, we commend the model to fellow researchers for further testing, development, and elaboration. Please, help yourselves.

REFERENCES

Adey, P. (1997). It all depends on the context, doesn't it? Searching for general, educable, dragons. *Studies in Science Education, 29*, 45-92.

Adey, P. (1993). *The King's-BP CASE INSET pack*. London: BP Educational Services.

Adey, P. (2004). Accelerating the Development of General Cognitive Processing. In A. Demetriou & A. Raftopoulos (Eds.), *Emergence and transformation in the mind: Modelling and measuring cognitive change*. Cambridge: Cambridge University Press.

Adey, P., Robertson, A., & Venville, G. (2001). *Let's Think!* Slough, UK: NFER-Nelson.

Adey, P., Robertson, A., & Venville, G. (2002). Effects of a cognitive stimulation programme on Year 1 pupils. *British Journal of Educational Psychology, 72*, 1-25.

Adey, P., & Shayer, M. (1993). An exploration of long-term far-transfer effects following an extended intervention programme in the high school science curriculum. *Cognition and Instruction, 11*(1), 1 - 29.

Adey, P., & Shayer, M. (1994). *Really Raising Standards: cognitive intervention and academic achievement*. London: Routledge.

Adey, P., Shayer, M., & Yates, C. (1993). *Naturwissenschaftlich denken* (H. A. Mund, Trans.). Aachen: Aachener Beiträge zur Pädagogik.

Adey, P., Shayer, M., & Yates, C. (2003). *Thinking Science Professional*. Cheltenham: Neslon Thornes.

Adhami, M., Johnson, D. C., & Shayer, M. (1998). *Thinking Mathematics: The Curriculum Materials of the CAME project*. London: Heinemann.

Atkinson, T., & Claxton, G. (2000). Introduction. In T. Atkinson & G. Claxton (Eds.), *The Intuitive Practitioner. On the value of not always knowing what one is doing*. Buckingham: Open University Press.

Avgitidou, S. (1997). Developing professional development in initial teacher training: a developmental model of student teachers' understanding of the relationship between theory and practice. Athens: EARLI.

Baird, J. R., Fensham, P. J., Gunstone, R. F., & White, R. T. (1991). The importance of reflection in improving science teaching and learning. *Journal of Research in Science Teaching, 28*(2), 163 - 182.

Bandura, A. (1997). *Self-efficacy: The excercise of control*. New York: W.H. Freeman.

Bell, B., & Gilbert, J. (1996). *Teacher Development: A Model for Science Education*. London: Falmer.

197

Blazely, L. D., Samnai, M., Rahayu, Y. S., & Purwati, R. (1996). *JSE Science: Diagnostic Survey* (10). Jakarta: Directorate of Secondary Education, Ministry of Education and Culture.

Bolam, R. (1982). *School Focused Inservice*. Oxford: Heinemann.

Borko, H., & Puttnam, R. T. (1995). Expanding a teacher's knowledge base: a cognitive psychological perspective on professional development. In T. R. Guskey & M. Hubermann (Eds.), *Professional Development in Education: New Paradigms and Practices*. New York: Teacher' College Press.

Brophy, J., & Good, T. (1986). Teacher behaviour and student achievement. In M. Wittrock (Ed.), *Handbook of Research on Teaching* (3rd ed., pp. 328-375). New York: Macmillan.

Brown, A. L. (1987). Metacognition, executive control, self-regulation and other more mysterious mechanisms. In R. Kluwe & F. Weinert (Eds.), *Metacognition, Motivation and Understanding* (pp. 65-116). London: Lawrence Erlbaum.

Brown, L., & Coles, A. (2000). Complex decision making in the classroom: the teacher as an intuitive practitioner. In T. Atkinson & G. Glaxton (Eds.), *The Intuitive Practitioner. On the value of not always knowing what one is doing.* Buckingham: Open University Press.

Brown, M., Askew, M., Baker, D., Denvir, H., & Millett, A. (1998). Is the national numeracy strategy research-based? *British Journal of Educational Studies, 46*, 362-385.

Brown, M., Askew, M., Millett, A., & Rhodes, V. (2003). The Key Role of Educational Research in the Development and Evaluation of the National Numeracy Strategy. *British Educational Research Journal, in press.*

Brown, M., Millett, A., Bibby, T., & Johnson, D. C. (2000). Turning our attention from the What to the How: the National Numeracy Strategy. *British Educational Research Journal, 26*(4), 457-471.

Calderhead, J. (1993). The contribution of research on teachers' thinking to the professional development of teachers. In C. Day & J. Calderhead & P. Denicolo (Eds.), *Research on teacher thinking*. London: Falmer Press.

Campbell, D., & Stanley, J. (1963). *Experimental and Quasi-experimental Designs for Research*. Chicago: Rand-McNally.

Choi, B.-S., & Han, H.-S. (2002). Cognitive acceleration of primary and middle school students through 'Thinking Science' activities. Harrogate: 10th International Conference on Thinking.

Cohen, L., & Manion, L. (1994). *Research Methods in Education* (4th ed.). London: Routledge.

Cooper, C., & Boyd, J. (1999). Creating sustained professional growth through collaborative reflection. In C. M. Brody & N. Davidson (Eds.), *Professional Development for Cooperative Learning* (pp. 49-62). Albany: State University of New York Press.

Cronbach, L., & Furby, L. (1970). How should we measure change, or should we? *Psychological Bulletin*(74), 68-80.

Cuban, D. (1988). A fundamental puzzle of school reform. *Phi Delta Kappan, 70*(5), 341-344.

De Corte, E. (1990). Towards powerful learning environments for the acquisition of problem solving skills. *European Journal of Psychology of Education, 5*(1), 5-19.

Department for Education and Employment. (2001). *Learning and teaching: A strategy for professional development*: DFEE Publications.

Department for Education and Skills. (2002). *Key Stage 3 National Strategy For the Foundation Subjects; Training Materials*. London: Department for Education and Skills.

Desforges, C., & Fox, R. (Eds.). (2002). *Teaching and Learning; The Essential Readings*. Oxford: Blackwell.

Desmione, L., Porter, A. C., Garet, M. S., Yoon, K. S., & Birman, B. F. (2002). Effects of professional development of teachers' instruction: results from a three year longitudinal study. *Educational Evaluation and Policy Analysis, 24*(2), 81-112.

Dillon, J., Osborne, J., Fairbrother, R., & Kurina, L. (2000). *A Study into the Views and Needs of Science Teachers in Primary and Secondary State Schools in England. Final Report to the Council for Science and Technology*. London: King's College London.

Donne, J. (1571-1613). Meditation XVII.

Doyle, W. (1977). Paradigms for research on teacher effectiveness. In L. Shulman (Ed.), *Review of Research in Education* (Vol. 5, pp. 163-198). Washington: AERA.

Driver, R., Squires, A., Rushworth, P., & Wood-Robinson, V. (1994). *Making sense of secondary science*. London: Routledge.

Dweck, C. S. (1991). *Self-theories and Goals: Their role in motivation, personality and development* (Vol. 38). Lincoln: University of Nebraska Press.

Earl, L., Watson, N., Levin, B., Leithwood, K., Fullan, M., & Torrance, N. (2003). *Watching and Learning 3: Final report of the external evaluation of England's National Literacy and Numeracy Strategies*. Nottingham: Department for Education and Skills.

Elmore, R. F. (1996). Getting to scale with good educational practice. *Harvard Educational Review, 66*(1), 1-26.

Egglestone, J. (1984). *An evaluation of the PKG Inservice-Onservice teacher education project*. Jakarta: Directorate of Secondary General Education, Ministry of Education.

Endler, L. C., & Bond, T. G. (2001). Cognitive development in a secondary science setting. *Research in Science Education, 30*(4), 403-416.

Eysenck, H. J. (1953). *Uses and Abuses of Psychology*. Harmondsworth: Penguin.

Fenstermacher, G. D. (1979). A philosophical consideration of recent research on teacher effectiveness. In L. Shulman (Ed.), *Review of Research in Education* (Vol. 6, pp. 163-198). Washington: AERA.

Forsman, J., Adey, P., & Barber, J. (1993). *A Thinking Science Curriculum.* Paper presented at the American Association for the Advancement of Science, Boston.

Fullan, M. (1995). The limits and potential of professional development. In T. R. Guskey & M. Hubermann (Eds.), *Professional Development in Education: New Paradigms and Practises.* New York: Teachers' College Press.

Fullan, M. (1999). *Change Forces- The Sequel.* London: Falmer.

Fullan, M. (2001). *Leading in a Culture of Change.* San Francisco: Jossey-Bass.

Fullan, M. G. (1982). *The Meaning of Educational Change.* London: Cassell.

Fullan, M. G., & Stiegelbauer, S. (1991). *The New Meaning of Educational Change.* London: Cassell.

Furlong, J. (2000). Intuition and the crisis in teacher professionalism. In T. Atkinson & G. Claxton (Eds.), *The Intuitive Practitioner. On the value of not always knowing what one is doing.* Buckingham: Open University Press.

Gage, N. L. (1978). *The Scientific Basis of the Art of Teaching.* New York: Teachers' College Press.

Gardner, P. L. (1974). Research on teacher effects: critique of a traditional paradigm. *British Journal of Educational Psychology, 44*(2), 123 - 130.

Garmston, R. J. (1987). How administrators support peer coaching. *Educational Leadership, 45*(2), 18-26.

Garmston, R. J., Linder, C., & Whitaker, J. (1993). Reflection on cognitive coaching. *Educational Leadership, 51*(2), 57-61.

General Teaching Council for England. (2003). *The Teachers' Professional Learning Framework.* London: GTC.

Gewirtz, S., & Ball, S. (1995). *Markets, Choice, and Equity in Education.* Buckingham: Open University Press.

Gilligan, Carol (1993) *In A Different Voice. London*; Harvard University Press

Gleick, J. (1999). *Faster.* New York: Pantheon.

Goddard, R. D., Hoy, W. K., & Hoy, A. W. (2000). Collective Teacher Efficacy: Its meaning, measure, and impact on student achievement. *American Educational Research Journal, 37*(2), 479-507.

Graham, J., Gough, B., & Beardsworth, R. (2000). *Partnerships in Continuing Professional Development.* London: Standing Committee for the Education and Training of Teachers.

Guskey, T. R. (1986). Staff development and the process of teacher change. *Educational Researcher, 15*(5), 5-12.

Guskey, T. R. (2000). *Evaluating Professional Development.* London: Sage.

Guskey, T. R., & Huberman, M. (1995). *Professional Development in Education: New Paradigms and Practises.* New York: Teachers' College Press.

Hall, G. E., & Loucks, S. F. (1977). A developmental model for determining whether the treatment is actually implemented. *American Educational Research Journal, 14*(3), 238-237.

Hargreaves, A. (1992). Cultures of teaching. In A. Hargreaves & M. Fullan (Eds.), *Understanding Teacher Development.* London: Cassell.

Hargreaves, A. (1994). *Changing Teachers, Changing Times: Teachers' work and culture in the postmodern age*. London: Cassell.

Hargreaves, D. (1997). In defence of research for evidence-based teaching: a rejoinder to Martyn Hammersley. *British Educational Research Journal, 23*(4), 405-419.

Hautamäki, J., Kuusela, J., & Wikström, J. (2002). CASE and CAME in Finland: "The second wave". Harrogate: 10th International Conference on Thinking.

Hennessy, G. M. (1999). Probing the dimensions of metacognition: implications for conceptual change teaching-learning. Boston: NARST Annual.

Hopkins, D. (Ed.). (1986). *Inservice Training and Educational Development: An International Survey*. Dover NH: Croom Helm.

Hopkins, D., & Lagerweij, N. (1996). The school improvement knowledge base. In D. Reynolds & R. Bollen & B. Creemers & D. Hopkins & L. Stoll & N. Lagerweij (Eds.), *Making God Schools*. London: Routledge.

Huberman. (1995). Professional careers and professional development, some interactions. In T. R. Guskey & M. Huberman (Eds.), *Professional Development in Education: New Paradigms and Practises*. New York: Teachers' College Press.

Huberman, A. M., & Miles, M. B. (1984). *Innovation Up Close: How School Improvement Works*. Newy York: Plenum.

Inhelder, B., & Piaget, J. (1958). *The Growth of Logical Thinking*. London: Routledge Kegan Paul.

Joyce, B. (1991). The Doors to School Improvement. *Educational Leadership*, 59-62.

Joyce, B., Calhoun, E., & Hopkins, D. (1999). *The New Structure of School Improvement*. Buckingham: Open University Press.

Joyce, B., & Showers, B. (1980). Improving inservice training; the messages of research. *Educational Leadership, 37*(5), 379-385.

Joyce, B., & Showers, B. (1982). The coaching of teaching. *Educational Leadership, 40*(1), 4-16.

Joyce, B., & Showers, B. (1988). *Student Achievement through Staff Development* (1st ed.). New York: Longman.

Joyce, B., & Showers, B. (1995). *Student Achievement through Staff Development* (2nd ed.). New York: Longman.

Joyce, B., & Weil, M. (1986). *Models of teaching* (3rd ed.). Englewood Cliffs NJ: Prentice-Hall.

Korthagen, F. A. J., & Kessels, J. P. A. M. (1999). Linking theory and practice: Changing the pedagogy of teacher education. *Educational researcher, 28*(4), 4-17.

Kuhn, D. (1999). A developmental model of critical thinking. *Educational Researcher, 28*(2), 16-26, 46.

Mahady, R., Wardani, I. G. A. K., Irianto, B., Somerset, A., & Nielson, D. (1996). *Secondary Education in Indonesia: Strengthening Teacher Competency and*

Student Learning (1a). Jakarta: Directorate of General Secondary Education, Department of Education and Culture.

McLaughlin, M. (1994). Strategic sites for teachers professional development. In P. Grimmett & J. Neufeld (Eds.), *Teacher Development and the Struggle for Authenticity: Professional Growth and Restructuring in a Context of Change* (pp. 31-51). New York: Teachers College Press.

McMahon, A. (2000). The development of professional intuition. In T. Atkinson & G. Glaxton (Eds.), *The Intuitive Practitioner. On the value of not always knowing what one is doing.* Buckingham: Open University Press.

Mevarech, Z. E. (1995). Teachers' paths on the way to and from the professional development forum. In T. R. Guskey & M. Hubermann (Eds.), *Professional Development in Education: New Paradigms and Practises.* New York: Teachers' College Press.

Miles, M. B., & Huberman, A. M. (1984). *Qualitative Data Analysis.* Newbury Park CA: Sage.

Mortimore, P., Sammons, P., Ecob, R., Stoll, L., & Lewis, D. (1988). *School Matters: The Junior Years.* Salisbury: Open Books.

Nam, J.-H., Choi, M.-H., Lee, S.-K., & Choi, B.-S. (2002). An analysis of problem solving process in the performance of CASE activities. Harrogate: 10th International Conference on Thinking.

Newman, D., Griffin, P., & Cole, M. (1989). *The construction zone: working for cognitive change in school.*: Cambridge University Press.

Office for Standards in Education. (2002a). *The National Literacy Strategy: the first four years* (HMI 555). London: OfSTED.

Office for Standards in Education. (2002b). *The National Numeracy Strategy: the first three years* (HMI 554). London: OfSTED.

Office for Standards in Education. (2003). *The Key Stage 3 Strategy: evaluation of the second year*: OfSTED.

Pfundt, H., & Duit, R. (1988). *Bibliography: students' alternative frameworks and science education* (5th ed.). Kiel: IPN.

Richardson, V. (1994). Conducting Research on Practice. *Educational Researcher, 23*(5), 5-10.

Robertson, A. (2002). Pupils' understanding of what helps them to learn. In M. Shayer & P. Adey (Eds.), *Learning Intelligence; Cognitive Acceleration Across the Curriculum* (pp. 51-64). Buckingham: Open University Press.

Rosenholtz, S. J. (1989). *Teachers' Workplace: The Social Organiazation of Schools.* New York: Longman.

Rutter, M. (1980). School effects on pupil progress - findings and policy implications. *Child Development, 54*(1), 1-29.

Sadtono, E., Handayani, & O'Reilly, M. (1996). *English Diagnostic Survey with Recommendations for Inservice Training Program for SLTP Teachers* (8a). Jakarta: Directorate of Secondary Education, Ministry of Education and Culture.

Schön, D. A. (1987). *Educating the Reflective Practitioner*. San Francisco: Jossey-Bass.

Shayer, M. (1996). *Long term effects of Cognitive Acceleration through Science Education on achievement: November 1996*: Centre for the Advancement of Thinking.

Shayer, M. (1999a). Cognitive Acceleration through Science Education II: its effect and scope. *International Journal of Science Education, 21*(8), 883-902.

Shayer, M. (1999b). *GCSE 1999: Added-value from schools adopting the CASE Intervention*. London: Centre for the Advancement of Thinking.

Shayer, M., & Adey, P. (1981). *Towards a Science of Science Teaching*. London: Heinemann.

Shayer, M., & Adey, P. (1992). Accelerating the development of formal thinking III: Testing the permanency of the effects. *Journal of Research in Science Teaching, 29*(10), 1101-1115.

Shayer, M., & Adey, P. (1993). Accelerating the development of formal operational thinking in high school pupils, IV: Three years on after a two-year intervention. *Journal of Research in Science Teaching, 30*(4), 351-366.

Shayer, M., & Adey, P. (Eds.). (2002). *Learning Intelligence: Cognitive Acceleration Across the Curriculum from 5 to 15 Years*. Milton Keynes: Open University Press.

Shayer, M., Küchemann, D., & Wylam, H. (1976). The distribution of Piagetian stages of thinking in British middle and secondary school children. *British Journal of Educational Psychology, 46*, 164-173.

Shayer, M., & Wylam, H. (1978). The distribution of Piagetian stages of thinking in British middle and secondary school children. II - 14- to 16- year olds and sex differentials. *British Journal of Educational Psychology*(48), 62-70.

Showers, B., & Joyce, B. (1996). The evolution of peer coaching. *Educational Leadership*, 12-16.

Shulman, L. (1987). Knowledge and teaching. *Harvard Educational Review, 57*, 1-22.

Somerset, A. (1996). *Junior Secondary School Mathematics: Diagnostic Survey of Basic Number Skills*. Jakarta: Directorate of General Secondary Education, Ministry of Education and Culture.

Stenhouse, L. (1975). *An introduction to curriculum research and development*. London: Heinemann Educational Books.

Stoll, L., & Fink, D. (1996). *Changing Our Schools: Linking School Effectiveness and School Improvement*. Buckingham: Open University Press.

Thair, M., & Treagust, D. (1997). A review of teacher development reforms in Indonesian secondary science: The effectiveness of practical work in biology. *Research in Science Education, 27*(4), 581-597.

Thair, M., & Treagust, D. (2003). A brief history of a science teacher professional development initiative in Indonesia and the implications for centralised teacher development. *International Journal of Educational Development, 23*(2), 201-213.

Thiessen, D. (1992). Classroom-based teacher development. In A. Hargreaves & M. Fullan (Eds.), *Understanding teacher development*. New York: Teachers College Press.

Tobin, K., Kahle, J. B., & Fraser, B. J. (Eds.). (1990). *Windows into Science Classrooms*. Lewis: The Falmer Press.

Toffler, A. (1970). *Future Shock*. New York: Bantam.

Tomlinson, P. (1998). Implicit learning and teacher preparation: potential implications of recent theory and research. Brighton: British Psychological Society Annual Conference.

Van den Berg, E. L., V. (1984). Science Teacher Diploma Programs in Indonesia. *Science Education, 68*(2), 195-203.

Venville, G. (2002). Enhancing the quality of thinking in Year 1 classes. In M. Shayer & P. Adey (Eds.), *Learning Intelligence; Cognitive Acceleration Across the Curriculum* (pp. 35-50). Buckingham: Open University Press.

Venville, G., Adey, P., Larkin, S., & Robertson, A. t. t. s. i. t. e. y. o. s. (2003). Fostering thinking through science in the early years of schooling. *International Journal of Science Education, in press*.

Vosniadou, S., Ioannides, C., Dimitrakopolou, A., & Papademetriou, E. (2001). Designing learning environments to promote conceptual change in science. *Learning and Instruction, 11*, 38-419.

Waring, M. (1979). *Social Pressures and Curriculum Innovation; A Study of the Nuffield Foundation Science Teaching Project*. London: Methuen.

Wenger, E. (1998). *Communities of practice: Learning, meaning, and identitiy*. Cambridge, UK: Cambridge University Press.

White, R. T., & Mitchell, I. J. (1994). Metacognition and the Quality of Learning. *Studies in Science Education, 23*, 21-37.

Wilson, S. M., & Berne, J. (1999). Teacher learning and the acquisition of professional knowledge: An examination of research on contemporary professional development. In A. Iran-Nejad & P. D. Pearson (Eds.), *Review of Research in Education* (Vol. 24). Washington DC: American Educational Research Association.

Worthen, B. R., & Sanders, J. R. (1987). *Educational Evaluation*. New York: Longman.

Wu, P. C. (1987). Teachers as staff developers: research, opinions, and cautions. *Journal of Staff Development, 8*, 4-6.

Yarrow, A., & Millwater, J. (1997). Evaluating effectiveness of a professional development course in supervising and mentoring. *British Journal of Inservice Education, 23*, 349-361.

INDEX